Springer Texts in Statistics

Springer Texts in Statistics

Braxton M. Alfred

Elements of Statistics for the Life and Social Sciences

With 18 Figures

Springer-Verlag
New York Berlin Heidelberg
London Paris Tokyo

Braxton M. Alfred
Department of Anthropology and Sociology
University of British Columbia
Vancouver, BC V6T 2B2
Canada

Editorial Board

Stephen Fienberg
Department of Statistics
Carnegie-Mellon University
Pittsburgh, PA 15213
USA

Ingram Olkin
Department of Statistics
Stanford University
Stanford, CA 94305
USA

AMS Classification: 62-01

Library of Congress Cataloging-in-Publication Data
Alfred, Braxton M.
    Elements of statistics for the life and social
sciences.
    (Springer texts in statistics)
    Bibliography: p.
    Includes index.
    1. Social sciences—Statistical methods.   2. Life
sciences—Statistical methods.   I. Title.   II. Series.
HA29.A53   1987   300′.1′5195   87-4964

Typeset by Asco Trade Typesetting Ltd., Hong Kong.
Printed and bound by R.R. Donnelley & Sons, Harrisonburg, Virginia.

9 8 7 6 5 4 3 2 1

ISBN-13: 978-1-4612-9138-1   e-ISBN-13: 978-1-4612-4744-9
DOI: 10.1007/978-1-4612-4744-9

Preface

This book was written to myself at about the time I began graduate studies in anthropology—the sort of thing a Samuel Beckett character might do. It is about the conduct of research. In a very real sense the purpose is partially to compensate for the inadequacies of my professors. Perhaps this is what education is about. The effort has not been an unqualified success, but it has been extremely gratifying.

I was trained in anthropology. After completing the Ph.D. I went to Stanford on a post-doctoral fellowship. At the time, this was a novelty and the department was not prepared for such a thing. To stay occupied I began attending lectures, seminars, and discussion groups in mathematics and statistics. This was about the luckiest choice I ever made. The excitement was easily as intense as that which I experienced upon encountering anthropology. On one occasion I innocently and independently proved a theorem that had first been done 2000 years earlier. It is currently used as an exercise in high school mathematics so it is neither difficult nor arcane. Learning all this did not tarnish my sense of discovery. (On reflection I am puzzled by my failure to have seen all this "beauty" when I was exposed to it as an undergraduate. The unparalleled excellence of the Stanford program was undoubtedly responsible for my belated conversion.)

Because it is the body of literature with which I am most familiar, examples have been drawn primarily from anthropology. The problems of this discipline are assuredly not unique to it. They are, on the contrary, common throughout the life and social sciences. The realization that there is no theory in anthropology may come as a surprise. There are, after all, several tons of books in most university libraries purporting to be anthropological theory. But the authors do not produce anything that is recognizable by other scientists as a theory (Alexander, 1981). Often abstruse language is called theoretical. There is never a dynamic system of equations expressing relationships among entities at a more basic level of reality. Occasionally one encounters the assertion that anthropology is not a science but an art. Leaving aside the undeniable fact that there is art in all science, notice that the name of the discipline means "science of man." This requires the development of theory with deductive power such that hypotheses may be produced rigorously. Then the hypotheses must be exposed to the most severe risk of rejection.

It seems reasonable to me that there should be little or no interest among students for techniques of testing the truth of gratuitous statements with no apparent origin. On the other hand, it also seems reasonable that the converse is true—the truth of statements generated by a theory should be of vital interest. If this is the case, then the only salvation for anthropology, the science, is in powerful rigorous theory.

Learning is often described in terms of directions: the process is either top-down, or bottom-up. The latter is illustrated by Robert Pirsig's *Zen and the Art of Motorcycle Maintenance*. The basic argument is that quantum mechanics theorists are produced from the ranks of motorcycle mechanics. The former is illustrated by Victor Weisskopf using quantum mechanics to explain why mountains and waves are as big as they are. I doubt that either approach can exist without the other. It is quite artificial even to characterize the process in an either/or manner. Individuals, philosophical traditions, academic departments, ... seem to have stable preferences, however. The dominant bias in anthropology in the 20th century has been for the empirical, bottom-up, approach. The expectation of this attitude is that truth will eventually emerge from data. Except in the trivial case of "what is, is true," this never works.

The bias here is a preference for a top-down approach. So at the outset I shall present the basic structure of science and place the conduct of anthropology within that larger tradition. The focus of attention then shifts to the problem of devising conditional statements such that there is a relatively rigorous connection between antecedent and consequent. And then finally we face the problem of determining whether the consequent, or its negation, occurred in a set of observations.

There is a great deal of repetition here. I make no apology for it. It is a pedagogic technique of great force and antiquity. Whether there is too much or too little is an empirical problem answerable only by the reader.

The formal requirements for reading and using this book are high school algebra and a calculator that will return natural logarithms. I have indicated the places where I expect the reader to accept a statement because its proof needs a higher level of mathematics. In the matter of less tangible requirements, there is an attribute called "mathematical maturity" which anyone can acquire in a fashion analogous to biological maturity. These two differ in that the former requires an act of will whereas the latter needs only survival. Otherwise exposure time is the critical factor.

It is useful to have or acquire some mathematical maturity. I hope that this book may contribute to the process for you. This truly is an overview, however. As such, greater demands are made on the student than is the case for a treatment which burrows deeper and deeper.

With two exceptions, all the computing here was done with a hand-held calculator or on a small microcomputer. The exceptions are Sections 5.4.1.1.4. and 5.4.2. As the techniques discussed there use iteration, I would still be working at it using a calculator. Admittedly, more complex problems certainly require more computational muscle, but I want you to realize that the essential ingredient in all successful and important research is the human mind. I tend to be suspicious of the attitude which sends students to the computer from the beginning. Typically one can, and so does, obtain highly complex results just because they are available. All too frequently there is no comprehension. One of the consequences of doing the work on a calculator is that the frequency of blundering errors increases. All the calculations have been checked repeatedly, and several classes of entering graduate students have verified each result. There are, undoubtedly, some errors produced by the accumulation of rounding error. These are rare, however. I trust that the quality control procedure used has made the other kinds of errors equally rare. Needless to say, I shall be grateful for your making me aware of those which have slipped through.

There is no set of problems for this edition, a deficiency which I shall rectify in the future. There are a number of completely worked examples (especially in Chapter 5), however. While these are no substitute for problems, they can assist in the process of acquiring concepts. You are urged to work through the examples, that is, to participate actively. There really is no alternative to participation. If you do it, the minimum result will be some enhanced "numeracy." If you do not do it, you should set things aside and return later.

While this is self-contained within the set bounds, I encourage you to pursue items in the Bibliography selectively. This book pretends to be neither comprehensive nor the ultimate authority on anything. There are, rather, two goals: (1) to sketch in the process of doing believable research, and (2) to provide enough detail for a few topics that you may use the recipes directly.

It is ironic that anthropology is becoming increasingly "provincial." Anthropologists are the only ones to accept what anthropologists say. Most of the life sciences share a similar problem. The main emphasis of the book will be on statistical tests of hypotheses, that is, deciding whether an experiment has or has not supported a hypothesis. But this is the very tag end of the entire research effort and so in the beginning, in the early parts, I shall consider the entire research framework in order to orient and motivate the statistical effort.

Anthropology and Science

A basic assumption here is that anthropology must be done scientifically. For now, this means, simply, that anthropologists must pay attention to the structure of argument, measurement, design and execution of research, and rules of evidence used by the larger scientific community. Science and revelation are the only sources of believable information. As there is no logic of revelation, the believability of revealed truth cannot be determined rationally. Revelation cannot be discussed, only accepted or rejected. At the most general level, science has two basic goals: (1) to discover the truth, and (2) to communicate it. Both are essential. Neither one alone is sufficient to the scientific enterprise. This entire work is concerned with the communication aspect of science. That is, the basic problem is how do you, when you know the truth, convince me? It is a requirement for participation in the scientific community that communication be part of the activity, and that this communication always be directed to some idealized, perfect, audience.

In order to satisfy this audience that the truth has been presented, there are certain reasonably well-defined procedural rules for the communication. These rules make this aspect rather simple by comparison to discovery. There is no recipe for the discovery of truth, but once it is in hand the communication is rather straightforward. That part of the literature which makes no effort to play by the rules of the game is nothing more than entertainment. Entertainment has neither inherent value nor logical structure—the Greek dramatists to the contrary notwithstanding. When science is abandoned, then rules of evidence which apply to scientific enquiry are abandoned and the believability of statements is entirely an emotional matter. Whether or not such statements have information content is purely incidental. Entertainment may, or may not, have value but it should be labelled as such in order that the required willing suspension of disbelief be in effect.

What This Book Is Not About

There is nothing in here about data collection. Recognizing that many beginning students need guidance, reference is made to Slater's (1978) recommendations to ethologists. It is assumed that in all contexts the investigator knows the appropriate techniques for obtaining the necessary data.

Also, it is not about classical experimental design. There is no expectation that the data are obtained in a laboratory environment where all relevant conditions are under tight experimenter control.

Nor is it about deterministic theory, that is, the assumption is explicitly made that theories of interest in anthropology are statistical, involving statistical causation, and that deterministic models of anthropological phenomena are necessarily inadequate. That statement is highly arbitrary. And in truth, I hope that it is wrong—life would be much simpler—but there is no way to resolve this issue at the moment. I am of the opinion that the current popularity of statistics in anthropology is due to ignorance of true dynamics.

Some Aspects of the Structure of Anthropology

Anthropology is one of the sciences that expects the entire research activity, from theory through experiment, to be done by either one person or a small team. The result is expensive. In the larger scientific context, of course, very few dollars are spent on anthropology. When the return is not informative —according to generally accepted rules of evidence—then any price is too much.

The result of expecting every anthropologist to produce a new "theoretical" development is a seeming unending stream of banality and/or scholasticism. Only a handful have ever had both the creative vision and technical skills to develop powerful theory. It is worth noting that none of these could have designed, executed, and analyzed the critical experiments.

This work is prompted by about 10 years of attempting to teach statistical methods in research. All too often students find anthropology by a kind of intellectual random walk. Subjects that include any mathematics tend to be rejected summarily. Eventually and inevitably, faculty come to resemble their students. Since students are highly adaptive—all that I have seen qualify as Primates (I think)—their general intransigence is the failure of the discipline (their instructors) to convince them of the relevance of the techniques. This is not a conscious conspiracy. The problem is much deeper. It resides in the professional reward structure. Individuals are often recognized proportionally to the weight of their contribution. "Why should I," the student legitimately asks, "bother learning this stuff when all I really need is 37 pounds of printed English?" This is the first question to be addressed later.

Journalism

The only thing that differentiates anthropology (or any science) from journalism is the quality of the theories. Few of us are blessed with the requisite gifts for producing these things. (A very effective way around this difficulty is to take a mathematician to lunch today.) The absence of powerful theory results in journalism. The common use of statistics under these circumstances has conditioned the displacement reaction of rejecting statistics. That is, when the house falls down, blame the hammer, not the carpenter.

The activity called numerical description is rather insidious because it is adorned with the mantle of technical respectability while being nothing more than description; it "embellishes gossip" (James, 1890) and foists the result off on the next generation of students as somehow more important than verbal description, a specious conclusion.

Two Cultures

C.P. Snow's concept of two cultures—arts and sciences—unable to communicate seems particularly appropriate for anthropology except that the partitions of anthropology are at least two-dimensional. At the risk of creating straw-men, one may easily recognize two subcultures within the discipline: (1) a literary group, and (2) a numerical group. The former is routinely concerned with "big" questions, torturous reasoning, baroque writing, and is mathematically and numerically illiterate. The latter directs attention almost exclusively to "things" (if you can't count it, it doesn't count) and procedures, and is illiterate mathematically and in the traditional linguistic sense. The main difference between anthropology and physics in this regard is that the two groups of physicists acknowledge their interdependence. In anthropology we are still involved in attempting to eliminate the other group.

Vancouver, B.C.   Braxton M. Alfred

Contents

CHAPTER 1

Introduction

This introductory chapter will outline, and summarize the sequel. Only the most rudimentary kind of measurement technique is assumed, that is, counted data are exclusively the subject of analysis. The reasons for this are legion but two important ones are: (1) All the necessary logic and decision concepts can be developed without diverting attention to matters that are of secondary importance in the research activity, and (2) historically, and in the present, this is by far the most common level of measurement in anthropology. Physical anthropologists count genes, cultural anthropologists count kinship systems, archaeologists count tool types, linguists count phonemes, etc. It is required only to know whether an observation is: a sickled or normal red cell, Hawaiian or other kinship, Folsom or other point, vowel or not, etc. A primitive level of measurement is adequate to support powerful and sensitive decision making.

Formal Argument in Anthropology

Considerable attention will be given to formally structured arguments. One justification for this is, as Whitehead said, that a formalism is useful because it gives you one less thing to worry about. Also throughout this discussion we will be concerned only with statements, which, by definition, are either true of false. It is important to realize that a statement may be either. We make no prior judgment on its truth or falsity. All that is required is that one be able to determine whether a particular statement is, in fact, true or false. A statement whose truth is unknown but can be determined is called a contingent statement.

The purpose of formal structure in science is to provide a good justification for believing that a statement is either true or false. Many anthropologists (and others) are committed to the proposition that statements are justified by the weight of evidence. In the absence of comprehensive meta-theory, enabling critical experiments, this condition will persist. Weight of evidence is adequate support for a decision only when it cannot be deferred, as in jurisprudence. (The phrase, unfortunately, lends itself to a literal interpretation.)

A major part of the justification of scientific statements resides in the structure of the argument. An argument is a set of statements which is divided into two parts—the premises and the intended conclusion. There may be any number of statements in the argument but typically there will be several premises and a single conclusion.

In the next chapter the structure of argument will be considered formally. The first two statements in the argument are the conditionals

if [THEORY] *then* [HYPOTHESIS]
if [HYPOTHESIS] *then* [OBSERVATION].

Note carefully that a hypothesis is a consequence of a theory. It is not—to quote an anthropologist—a guess. The capricious formulation of "hypotheses" is responsible for much of the shoddy work that one encounters in the literature. Notice also that a hypothesis is logically justified when exhibited as the necessary conclusion of a formal argument. This produces deductive validity. An argument is said to be deductively valid when the truth of the premises makes it impossible that the conclusion is false. Deductive validity, however, is insufficient for the justification of scientific hypotheses. A deductive conclusion merely re-states part of the information given in the premises. Scientific reasoning cannot be adequately represented by deductive argument because scientific conclusions contain more information than any premises one might use to justify them. Arguments of this sort are called inductive. An argument is said to be a good inductive argument when the truth of its premises guarantee a high probability for the truth of its conclusion.

The Place of Statistics in Research

The theory being investigated completely determines the importance, or worth, of any statistic. If the theory is important then the techniques allowing one to decide whether to accept or reject are important. Often, but not always, the techniques are statistical. If the theory is silly or trivial then nothing can salvage it—not mountains of data processed by high powered computers using the most sophisticated statistical methods.

There are three basic uses of statistics in the research environment—description, hypothesis testing, and estimation. There is a logical sequence to these: (1) "what is out there?", (2) "why is it the way it is?", and then (3) "what are the (numerical) values of the model parameters?" Research activity is considered to begin with the "why?" question. This is the phase which involves examining the implications of the theory developed to explain the observation obtained by asking the "what?" question. When a particular theory has survived the rigors of repeated attempts to falsify it, activity and interest shift to the problem of determining the values of the parameters.

For example, it has been observed (description) that the frequency of the sickle-cell trait in the U.S. is low among the young and high among the old (*sic*). The reason for this (theory) is that differential fertility favoring the normal hemoglobin phenotype is driving this evolutionary phenomenon. If this should be supported then we may be interested in the magnitude (estimation) of the fertility differential. Frequently researchers attempt to combine the test of a hypothesis and the estimation of parameters. But, as we shall see later, these two activities are fundamentally different—the goals, and hence the requirements are not the same. So it is not, in general, possible to do both things with the same project.

I wish also to emphasize the point that numerical results are not in any sense substitutable for careful thought. Numbers do not produce value. The widespread availability of muscular number-crunching computers has had the untoward effect of yielding power to the sorcerer's apprentice. Statistics and computers must support the research activity, not motivate it. If the theory is elegant and important, it may be adequately tested by a single observation. If it is clumsy, crude, and trivial you will not live long enough to collect and process the data to transform it.

What Is to Come?

In the next chapter, the propositional analysis of four common argument structures is presented. The goal of such an analysis is to determine the validity of an argument. The mechanism used is the truth table. Because invalid arguments produce false conclusions under some conditions, only valid arguments are acceptable in science. If the terms of the conditional are as above— HYPOTHESIS and OBSERVATION—then HYPOTHESIS only may be the antecedent. (A commonly encountered variation is to insert OBSERVA-TION as the antecedent. This will be considered later.) The result of the analysis produces the well known—Aristotle did it first—if non-intuitive conclusion that the only valid argument available to science is one called "Deny the Consequent." This condition produces the apparently bizarre requirement that if the predicted OBSERVATION is false then so is the HYPOTHESIS. Chapter 2 concludes with some logical pathologies in order to establish that valid arguments may produced absurd conclusions. So validity is necessary to, but not sufficient for, good science.

Chapter 3 addresses the matters of inductive argument, the scientific program, and the conditions necessary for a good test of a HYPOTHESIS. A deductive argument preserves truth—the conclusion is necessarily true—but cannot produce novelty. Without the ability to make novel assertions, there would be no science. An inductive argument facilitates novelty, but the conclusion can be only probabalistically true at best. Inductive argument is, therefore, an essential ingredient of science, the truth of conclusions is always only probable. The scientific program, then is a set of statements, which includes an inductive argument, leading to a conclusion which declares that the theory is probably true. The two conditions of a good test must be satisfied before the argument is constructed. These relate to the connection between HYPOTHESIS and OBSERVATION. The conditions are

(1) *if* [HYPOTHESIS] *then* [OBSERVATION], and
(2) *if* [*not* HYPOTHESIS] *then* [*not* OBSERVATION].

Condition 1 will be satisfied if OBSERVATION is a deductive consequence of HYPOTHESIS. Satisfying condition 2 is the most difficult step in the entire enterprise. It is a rare event.

In Chapter 4 the problem of deducing OBSERVATION from HYPOTHESIS is considered. Some basic principles of probability theory are presented initially and then the matter of formalizing the theory is discussed from the perspective of several kinds of well known models. This is the subject of a very large body of literature in mathematics. The brief survey there is intended to motivate intuition. Two kinds of stochastic models are presented —the Poisson and Markov. Also there is a model of structure in a series of events. The application is to choreography in social behavior. And finally, due to the breadth of applications (especially in evolutionary theory) there is an introductory description of some small games.

Chapter 5 is the primary goal of this effort. The scientific process culminates ultimately in the test of a hypothesis. In this chapter are presented techniques for testing a variety of hypotheses with frequency, categorical, data only. The variety which is accessible depends on the complexity of the structure of the observations. For example, observations of age and sex are more complex than observations of sex alone. While it is easily possible today to fit models, evaluate hypotheses, to very complex data, this chapter terminates with a discussion of a four dimensional structure. The reason for this arbitrary limit is that the results must be interpreted. When a hypothesis is very precise then complexity is not a serious matter. The model specified by the hypothesis is tested for for goodness of fit to data and is either rejected or not. It is frequently the case in the life and social sciences that data are also used in an exploratory manner. (OBSERVATION is inserted as the antecedent in the conditional statement.) This process may result in observing a good fit for an unanticipated model. If this is to be communicated to other scientists it must be fully interpreted. Interpreting a complex model for which there is no prior hypothesis may well be impossible.

In the summary, each of these topics is considered to be a step in a process. It is quite artificial to exhibit the sequence as if it were linear proceeding from rigorous logic, to model of the hypothesis, and finally to the experimental test of the hypothesis. The human mind does not operate this way and so it is doubtful if any project has ever followed this trajectory. These are, however, the *ad hoc* minimal components of believable research.

CHAPTER 2

Some Elementary Principles of Deductive Argument

The burden of communication in the arts is shared between artist and audience. This allows the artist, regardless of his effort, to be "misunderstood." In science the burden is totally on the initiator, the scientist. Clarity of logic is, therefore, an essential ingredient. Formalism is a useful, and in most cases of interest, necessary, part of the process.

Statements, Arguments

The statement is the building block. A statement is a declarative sentence which is either true or false. It cannot be both true and false; nor can it be neither true nor false. It must be exactly one or the other.

Scientific arguments are conditional arguments. A conditional argument includes some conditional statements. A conditional statement is characterized by an *if* [·] *then* [·] structure. The statement following the *if* is called the antecedent and that which follows the *then* is called the consequent.

Conventions

We shall adopt the following notational conventions: [P] will symbolize the antecedent and [Q] the consequent. The standard tool for determining the truth of an argument or compound statement of any kind is the truth table. In the early parts of this chapter we shall consider some of the basic principles of propositional logic and truth tables.

Since all statements are either TRUE or FALSE, each has a property called the truth value. (Upper case will be used to refer to the truth value of a specific statement.) Compound statements are conjunctions of simple statements. The truth value of any compound statement is determined entirely by the truth values of its components. There is a very small number of ways of connecting statements. In fact, in this section we will consider only four. They are called conjunction, disjunction, negation, and conditional. In common language, conjunction is indicated by the word *and*; disjunction is indicated by *or*; and negation by *not*; conditional is indicated by *if* [·] *then* [·].

A further convention involves the construction of the truth tables themselves. The leftmost columns will always be dedicated to the truth values of the simple statements. Note that the number of lines in a truth table is completely determined by the number of simple statements which are involved. Each line is a different combination of the truth values of the simple statements. If there is but a single statement, there are two lines in the truth table because the statement can be only true or false. If there are two simple statements, each of which may be either true or false, then there will be four lines in the truth table. Extending this we note that if there are three simple statements, each of which may be true or false, there will be eight lines in the truth table. In general, when there are n simple statements, there will be 2^n lines in the truth table.

2.1. Common Connectives for Statements

2.1.1. Conjunction and Disjunction

The truth values of conjunctions and disjunctions for two simple statements under all possible values of the simple statements are presented in Table 2.1. A bit of reflection will convince you that these values for the connectives, that is for the compound statements, are intuitively quite plausible. For example, conjunction is true only if both of the simple statements are true. It is not the case that the conjunction can be true when either or both of the simple statements are false. And note that just the reverse is true for disjunction. That is, when either or both of its simple statements are true the disjunction itself is true, but when both of the simple statements are false, the

Table 2.1. Truth Values of Conjunction and Disjunction

Simple statements		Conjunction	Disjunction
P	Q	P *and* Q	P *or* Q
T	T	T	T
T	F	F	T
F	T	F	T
F	F	F	F

disjunction is also false. No truth table will be presented for the connective negation because it seems straightforward: if [P] is TRUE, [*not* P] must be FALSE. And conversely if [P] is FALSE, then [*not* P] is TRUE. That is to say, negation simply reverses the truth value.

2.1.2. The Conditional

Science depends very heavily on the use of conditional statements of the form *if* [P] *then* [Q]. The truth table for the conditional is presented in Table 2.2. Consider the left-most three columns of this table. The first two lines are sensible: when the premise [P] is TRUE, then if the conclusion is TRUE, the conditional statement is TRUE, and when the premise is TRUE and the conclusion FALSE, the conditional statement is also FALSE.

False Premises

The last two lines defeat common sense and intuition abandons us entirely. For example, what are we to make of a statement which has a false premise and a true conclusion? What should we have for the truth value of the conditional? The value TRUE has been assigned to this particular construction as it has to the last line of the truth table. That is, when the premise is false and the conclusion is also false then the conditional statement *if* [P] *then* [Q] is true. No further motivation can be given to elaborate the truth values which have been assigned to the last two lines of this particular truth table. However, you should be aware that these values are not arbitrary and logicians are able to show that, if any other values are assigned to these lines of the table, then some totally unacceptable kinds of statements result. Fortunately the last two lines of this table need not deter us as they do not enter science. Science is only concerned with arguments which have true premises.

Functional Equivalents

Perhaps a bit of interpretive solace can be derived from a consideration of the last two columns of Table 2.2 where some functional equivalents of the conditional statements are presented. For example, the statement (*not* (P *and*

Table 2.2. Truth Values of the Conditional

P	Q	*if* P *then* Q	*not* (P *and not* Q)	*not* P *or* Q
T	T	T	T	T
T	F	F	F	F
F	T	T	T	T
F	F	T	T	T

not Q)) produces exactly the same truth table as the conditional statement. Also the statement (*not* P *or* Q) results in an identical truth table. As functional equivalents, these constructions may be substituted for the conditional wherever it is useful.

2.2. Argument

Valid Argument

Here we consider the validity of some common argument structures. A scientific argument always has three parts or constituent components: (1) a conditional statement, (2) an intermediate conclusion, and (3) an ultimate conclusion.

(1) *if* [P] *then* [Q]
(2) [intermediate conclusion]
(3) [ultimate conclusion].

In a formal argument of this sort the conditional statement and the intermediate conclusion are the premises of the argument. For our purposes the intermediate conclusion is that which is subject to observation, and the ultimate conclusion is the intended result of the argument.

Definition. An argument is valid when the conjunction of all constituent components is true.

Note that this definition means that: (1) both components of the conditional must be true, (2) the intermediate conclusion must be true, and (3) the ultimate conclusion must be true. Then the argument is valid.

2.2.1. Affirm the Antecedent

Consider as a first structure of argument

if [P] *then* [Q]
[P]
thus [Q].

This is read as *if* [P] *then* [Q], [P] is TRUE, therefore [Q] must be TRUE. In Table 2.3. the truth table for this argument structure is presented. Note from the heading of this table that the argument is sufficiently common that it has a name. It's called Affirming the Antecedent. Consider the third and fourth columns—*if* [P] *then* [Q] *and* [P]. Form the conjunction for each line of this truth table and produce column 5, (*if* [P] *then* [Q]) *and* [P], of the

Table 2.3. Truth Table for "Affirm the Antecedent"

Simple statement P Q	Conditional *if* P *then* Q	Intermediate conclusion (P)	(*if* P *then* Q) *and* P	Intended result (Q)	Ultimate conclusion
T T	T	T	T	T	T
T F	F	T	F	F	F
F T	T	F	F	T	F
F F	T	F	F	F	F

Table 2.4. Truth Table for "Deny the Consequent"

Simple statement P Q	Conditional *if* P *then* Q	Intermediate conclusion (*not* Q)	(*if* P *then* Q) *and not* Q	Intended result (*not* P)	Ultimate conclusion
T T	T	F	F	F	F
T F	F	T	F	F	F
F T	T	F	F	T	F
F F	T	T	T	T	T

table. The premises are jointly true only in the first row of this column. Now form the conjunction of the joint premises and the intended result producing column 7, the ultimate conclusion. In all cases involving true premises (row 1), the ultimate conclusion is true. An argument is valid if the conclusion is true when the premises are. This argument is valid.

2.2.2. Deny the Consequent

Let us turn now to a second common argument structure which also has acquired a name due to frequent usage. It's called Denying the Consequent, and its structure is presented as

if [P] *then* [Q]
not [Q]
thus [*not* P].

The truth table for this argument is presented in Table 2.4. Looking at this table and forming the conjunction of *if* [P] *then* [Q], and *not* [Q], for each row, we observe that the only time the premises are jointly true is in line 4. That is, the conditional is TRUE, and the intermediate conclusion *not* [Q] is also TRUE. It may seem exceedingly strange to produce a valid argument when both of the simple statements involved in the conditional statement are themselves false. This structure is basic to all science, however, and consequently we will have much more to say about it later.

Table 2.5. Truth Table for "Deny the Antecedent"

Simple statement P	Q	Conditional if P then Q	Intermediate conclusion (not P)	(if P then Q) and not P	Intended result (not Q)	Ultimate conclusion
T	T	T	F	F	F	F
T	F	F	F	F	T	F
F	T	T	T	T	F	F
F	F	T	T	T	T	T

2.2.3. Deny the Antecedent

A third argument structure is called Deny the Antecedent. Its structure is presented as

> if [P] then [Q]
> not [P]
> thus [not Q]

and its truth table in Table 2.5.

Only lines 3 and 4 are of concern in the determination of the validity of the argument. The conjunction of premises is true in both cases. Consider line 3. Notice that the truth value of the conditional is TRUE, as is that of the intermediate conclusion, *not* [P]. The conjunction of the conditional and the intermediate conclusion then, yields the value true. However, refer back to the truth value of the simple statement [Q]. In line 3, [Q] has a truth value of TRUE so *not* [Q] is FALSE. Therefore, this particular argument is invalid, and it is invalid specifically because we have an instance wherein both of the premises of the argument, that is, the conditional statement and the intermediate conclusion are TRUE, but the conclusion of the argument is FALSE.

2.2.4. Affirm the Consequent

And finally let's consider the argument structure called Affirm the Consequent. The structure is presented as

> if [P] then [Q]
> [Q]
> thus [P]

and the truth table is presented in Table 2.6.

Consider line 3. There both the conditional statement and the intermediate conclusion, [Q], are TRUE. Their conjunction, then, is TRUE. However, refer back to the truth value of the simple statement, [P], and observe that it (line 3) is FALSE. This argument structure, therefore, is invalid and specifically again because we have an instance wherein the premises of the argument are both TRUE and the conclusion FALSE. Even though invalid, this argument

Table 2.6. Truth Table for "Affirm the Consequent"

Simple statement P Q	Conditional *if* P *then* Q	Intermediate conclusion (Q)	(*if* P *then* Q) *and* Q	Intended result (P)	Ultimate conclusion
T T	T	T	T	T	T
T F	F	F	F	T	F
F T	T	T	T	F	F
F F	T	F	F	F	F

has a strong appeal to intuition. After all, if a theory predicts something which is then observed to be true, surely the theory is supported somehow. In fact Polya (1954) enshrines this as a principle:

> The verification of a consequence renders a conjecture more credible.

But, while credibility may be enough to sustain an interest in the theory, it is not a valid structure and, so, has the potential of producing erroneous statements. Consider the following:

> "Sire," he said to him, "I beg that you will excuse my asking you a question—"
> "I order you to ask me a question," the king hastened to assure him.
> "Sire—over what do you rule?"
> "Over everything," said the king, with magnificent simplicity.
> "Over everything?"
> The king made a gesture, which took in his planet, the other planets, and all the stars ....
> "And the stars obey you?"
> "Certainly they do," the kind said. "They obey instantly. I do not permit insubordination ...."
> "If I ordered a general to fly from one flower to another like a butterfly ... which one of us would be in the wrong?" the king demanded. "The general, or myself?"
> "You," said the little prince firmly.
> "Exactly. One must require from each one the duty which each one can perform," the king went on. "Accepted authority rests first of all on reason. If you ordered your people to go and throw themselves into the sea, they would rise up in revolution. I have the right to require obedience because my orders are reasonable."

> Antoine de Saint-Exupery
> *The Little Prince*

Define

[P]: "I am king of the universe."
[Q]: "The stars and planets obey me."

Then the king's argument is

if [P] *then* [Q]
[Q]
thus [P].

2.2.5. Decomposition of Arguments

Consider the statement (*not* ([P] *and not* [Q])). Let us determine the truth table for the statement.

[P]	[Q]	(3) *not*	(1) ([P]	(2) *and*	(1) *not* [Q])
T	T	T	T	F	F
T	F	F	T	T	T
F	T	T	F	F	F
F	F	T	F	F	T

In order to process the statement it was necessary to perform operations sequentially—the sequence of the operation is indicated above each column. First rewrite the statement to the right such that each variable and operator heads a new column. Since the operations, [P] and *not* [Q], are performed immediately, these truth values are entered and these columns are first in sequence. It is important to evaluate statements within parentheses before attempting higher level relationships. So the next step is to enter the truth table for the conjunction of [P] and *not* [Q] under the *and* within the parentheses. This column is the second, (2), step. The truth table in this column gives the value of the entire parenthesized statement so that all that remains to do is negate it and enter this as (3). This column is the truth table for the entire statement. As a point of interest, you should compare this result with column 4 of Table 2.2.

The order of processing is determined by the inner parentheses. Note that if they are removed the statement becomes (*not* [P] *and not* [Q]) which is quite a different statement.

Now reconsider the argument, affirm the consequent

> First premise: *not* [P] *or* [Q]
> Second premise: [Q]
> Intended result: *not* [P].

We may write this as

$$(((not\ [P]\ or\ [Q])\ and\ [Q])\ and\ not\ [P])$$

and create the truth table

[P]	[Q]	(1) (((*not* [P]	(2) *or*	(1) [Q])	(3) *and*	(1) [Q])	(4) *and*	(1) *not* [P])
T	T	F	T	T	T	T	F	F
T	F	F	F	F	F	F	F	F
F	T	T	T	T	T	T	T	T
F	F	T	T	F	F	F	F	T

Note that the argument is not valid because there is an instance of all premises being true but the intended result, (4), is false (line 1). More importantly, note how the argument was structured with parentheses so that its truth table could be evaluated. Also note how the parentheses determine the processing sequence.

2.2.6. Paradox

Lest you be tempted to think that all problems are, in principle, logical, here you will see only a small set of the "pathologies," otherwise known as paradox. Consider the argument

Premise: [P]
Intended result: *if* [Q] *then* [P].

Form the expression

([P] *and* (*if* [Q] *then* [P]))

and evaluate its truth table in Table 2.7. Notice that the conjunction of the single premise and the intended result is true when the premise is true so the argument is valid. As both [P] and [Q] are completely general with no constraints except that they be statements, this result demonstrates (proves) that a true statement is validly implied by any statement whether true or false.

Next we analyze

Premise: *not* [P]
Intended result: *if* [P] *then* [Q].

Form the expression

(*not* [P] *and* (*if* [P] *then* [Q]))

from which the truth table is constructed in Table 2.8. Since the conjunction of the single premise and the intended result is true when the premise is, the argument is valid. So we may conclude that any statement is validly implied by a false statement (see lines 3 and 4, column (2)).

Table 2.7. True Propositions Are Implied by Any Proposition

P	Q	(1) (P	(2) *and*	(1) (*if* Q *then* P))
T	T	T	T	T
T	F	T	T	T
F	T	F	F	F
F	F	F	F	T

Table 2.8. False Propositions Imply Anything

P	Q	(1) (*not* P	(2) *and*	(1) (*if* P *then* Q))
T	T	F	F	T
T	F	F	F	F
F	T	T	T	T
F	F	T	T	T

Table 2.9. Anything Is Implied by a Contradiction

P	Q	(1) ((P	(2) *and*	(1) *not* P)	(3) *and*	(1) Q)
T	T	T	F	F	F	T
T	F	T	F	F	F	F
F	T	F	F	T	F	T
F	F	F	F	T	F	F

Next we consider

        Premise: [P] *and not* [P]
        Intended result: [Q].

Create the expression

        (([P] *and not* [P]) *and* [Q])

and produce the truth table in Table 2.9. This example illustrates an important fact about validity and provides a sufficient reason for our restricting attention to arguments with true premises. This is an instance of our being unable to assert that the argument is invalid since there is no case of the premises being true and the conclusion false. We may, therefore, conclude that anything is implied by a contradiction.

    Next we consider

        Premise: [Q]
        Intended result: [P] *or not* [P].

Form the expression

        ([Q] *and* ([P] *or not* [P]))

and the truth table in Table 2.10.

Table 2.10. A Tautology May Be Validly
Inferred from Any Proposition

P	Q	(1) (Q	(3) and	(1) (P	(2) or	(1) not P))
T	T	T	T	T	T	F
T	F	F	F	T	T	F
F	T	T	T	F	T	T
F	F	F	F	F	T	T

Notice that when the premise is true so is the conclusion and we conclude that the argument is valid. From that we may infer that a tautology is validly implied by any proposition.

These examples, while curious, and I trust, disturbing, are not true paradoxes. Usually the conditions of logical paradox are: (1) self-reference, (2) contradiction, and (3) vicious circle (Hughes, 1975). For example:

1. This sentence is false.
2. Socrates: What Plato is about to say is false.
   Plato: Socrates has just spoken truly.
3. Hempel's Paradox of the crow seems to go to the heart of the scientific method. It is cast as follows:

A scientist wishes to investigate the hypothesis "All crows are black." He begins examining crows and the more black ones he finds, the more probable, he thinks, the hypothesis. Hempel proved that a purple cow would be a confirming instance of the hypothesis that all crows are black. Let:

[P]: crow
[Q]: black.

By hypothesis *if* [P] *then* [Q]. This statement can be shown to be equivalent to 'if not black then not crow' that is *if not* [Q] *then not* [P]. So a confirming observation for the second statement is a confirming observation of the first. Any non-black observation which is non-crow is confirmation of all crows are black! Clearly a purple cow is non-black and non-crow so it confirms the hypothesis.

2.2.7. Summary of Deductive Propositional Logic

Up to this point we have been considering deductive logic exclusively and it is time to recognize that this is not adequate for the development of a science. Later when considering induction we will have occasion to develop this more completely.

Table 2.11. Truth Table Summary of Argument Types

Simple statement P Q	Conditional *if* P *then* Q	Conjunction of premises			
		Affirm antecedent (P)	Deny consequent (*not* Q)	Deny antecedent (*not* P)	Affirm consequent (Q)
T T	T	T	F	F	T
T F	F	F	F	F	F
F T	T	F	F	T	T
F F	T	F	T	T	F

In Table 2.11. is presented the conjunction of premises for each of the argument types described in this chapter.

Since the intended result is not specified, validity is not determined.

CHAPTER 3

The Logic of Scientific Argument

Science and deductive logic are not equivalent. In this chapter the logical basis of science is presented. A general feature of all deductively valid arguments is that the conclusion contains no more information than that which is given in the premises. The conclusion simply restates the information in the premises. Giere (1984), on whom this disucssion depends heavily, defines a good inductive argument as

Definition. An argument is a good inductive argument if and only if the truth of its premises guarantee an appropriately high probability for the truth of its conclusion.

The main difference between good inductive and good deductive argument resides in the truth value of the conclusion. A deductively valid argument is necessarily true. A valid inductive argument is probably true. The distinguishing feature of an inductive argument is that the conclusion contains information which is not present in the premises, so, by definition, the conclusion cannot ever be deductively true. Furthermore, inductive arguments do not preserve truth; that is, it is possible for a good inductive argument to have a false conclusion even though all of its premises are true.

Theory and Hypothesis

It is probably also clear by now that the premises in an inductive argument are related to scientific theory in some way. It is useful at this point to distinguish a theory from a theoretical hypothesis.

Two statements given in the introduction are repeated here for reference.

1. *if* [THEORY] *then* [HYPOTHESIS]
2. *if* [HYPOTHESIS] *then* [OBSERVATION].

THEORY is a definition of a natural system, and HYPOTHESIS is a deductive consequence of the definition. Note that theories have the form of a definition and, so, say nothing about the world. An implied, but unstated, assertion allowing the second statement is that a specific real system is of the type that is defined by THEORY. This is the creative, insightful, step in science. There is no logic to support it. It is done by convention and expert prior knowledge of the real system. This is where the connection between the abstract, symbolic, HYPOTHESIS and the physical system is made. The strength of the second statement above comes entirely from the justification of this connection. Strong justification makes the second statement strong, and conversely. In the following discussion it will be assumed that this justification is strong. Ordinarily this will mean that the components, language, and forces of THEORY are those "customarily" used to describe the real system. Since it is not necessary that good theories give precise definitions of their key concepts, HYPOTHESIS in the first statement, the theoretical hypothesis, is justified by showing that it is the conclusion of an argument in which some other statements are premises. The transformation of the theoretical hypothesis into a research hypothesis, HYPOTHESIS in the second statement, is justified by the assertion that THEORY is a model of a real system.

The structural statements above are modified as

1.   *if* [THEORY] *then* [theoretical HYPOTHESIS]
1a. [THEORY is a model of the real system]
2.   *if* [research HYPOTHESIS] *then* [PREDICTION]

where PREDICTION is an OBSERVATION that should be made if THEORY is an accurate model. PREDICTION is an expected OBSERVATION.

Now consider the basic program of science.

3.1. The Program of Science

In this section is a version of the model of science. It is based on the argument structure of Section 2.2.2., deny the consequent. Recall that this is the only valid structure which allows the prediction to fill the role of logical consequent. Also expectably the model accommodates the definition of a good inductive argument.

1. Initiating premise:
   if [THEORY] *then* [theoretical HYPOTHESIS]
2. Transforming premise:
   [THEORY is a model of the real system]
3. First premise:
   if [*not* research HYPOTHESIS] *then* [probably *not* PREDICTION]

4. Second premise:
    [PREDICTION is probably TRUE]
5. Intermediate conclusion:
    thus [research HYPOTHESIS is probably TRUE]
6. Ultimate conclusion:
    thus [THEORY is probably TRUE]

Components of the Program

Before stating the model symbolically consider the components. Line 1, the initiating premise, is included as a reminder that HYPOTHESIS has a specific origin in the theory and does not exist independently. Lines, 3, 4, and 5 constitute a denial of the consequent argument. Notice that the antecedent is itself the negated hypothesis. In line 4 PREDICTION is observed, and line 5 is the valid conclusion—the negation of the antecedent. Line 6 follows only because of line 2.

Notice that the only inductive step in this scientific program is in line 4 of the program.

The initiating premise (line 1) in the argument is responsible for the vast majority of the scientific literature. It is necessary initially that one be able to show that a particular prediction (theoretical HYPOTHESIS) is a necessary consequence of THEORY. Additionally one must show that without the theory, the hypothesis does not result.

Line 2 establishes the connection of the theory to the realm of observation. Since it is difficult, if not impossible, as well as intellectually unsatisfying, to observe that a prediction did not occur, line 3 asserts the negation of the qualified line 1: if research HYPOTHESIS is not TRUE then PREDICTION will probably not be observed.

Now turn your attention to the second premise (line 4) in the argument, that is, the prediction has turned out to be (probably) true. This is the equivalent of denying the consequent in line 3 and so the basic structure of the argument is valid. It should be pointed out in passing that this second premise is the primary subject of most of this work, that is, the mechanisms and procedures by which one decides whether or not a particular prediction has in fact turned out to be (probably) true.

Recall that the prediction was not stated initially in absolute or precise terms. It was, on the contrary, stated (line 3) in terms of if the hypothesis is true then very probably the prediction is true. The determination of whether or not the prediction has been observed in the world is frequently non-trivial, and it constitutes the subject matter of that body of mathematics known as statistics. Note, however, that statistics alone is insufficient to determine whether the prediction has been observed. One also needs to design an experiment and/or other observational techniques, make measurements, and process the data.

The intermediate conclusion is stated in terms of the negation of the antecedent in the conditional statement (line 3). This follows because the prediction was observed to be true and as the consequent in the first premise it was stated in the negative.

Symbolic Rendering of the Program

This scientific program produces the result that when the prediction is true, then only by coincidence can the hypothesis be false. Notice that in recognizing the possibility that the hypothesis can be false under these conditions is in the nature of an inductive argument. This is an unlikely event when this program is satisfied. The program is typically shortened as

> *if not* [H] *then* probably *not* [P]
> [P]
> *thus* [H].

A Common Invalid Argument

This shortened form facilitates the perception of an invalid argument which is pervasive throughout much of the social and life sciences. Consider the argument

> *if* [H] *then* [P]
> [P]   ['affirm the consequent']
> *thus* [H].

This argument is: *if* [antecedent] *then* [consequent], the [consequent] is TRUE, *thus* [antecedent is TRUE]. This is an affirmation of the consequent and has been known to be an invalid argument structure for at least 2000 years. Even though invalid, the argument does seem to add "credibility" to the hypothesis (Polya, *op cit*). This, however, is outside of the scientific program.

3.2. Elements of a Good Test

Notice first of all that the reason for the test is to obtain the best possible judgment of the truth value of the theoretical hypothesis, not the accuracy of the prediction. This point establishes that we are not concerned with knowing how much of something is true, but only with whether the hypothesis is true. This has important implications for the design of the experiment. Secondly, note that the prediction depends upon the hypothesis through a deductive argument. The prediction is deduced from premises that include the hypothesis and could not be deduced from the remaining premises if the hypothesis were not present. In the event that HYPOTHESIS is TRUE but the prediction fails, it is the case that the prediction does not follow deductively from the hypothesis.

A condition not listed below is that the prediction must be verifiable. It must be possible to determine, when the experiment is completed, whether the prediction occurred or not. Some events are easier than others to verify. For example, a returning comet is easily observed. It is considerably more difficult to determine whether the mating of heterozygotes produces offspring phenotypes in the ratio of 3 : 1 dominants to recessives. And it seems to be nearly impossible to determine whether fraternal polyandry increases male fitness (Beall and Goldstein, 1981). At the risk of stating the obvious, if a hypothesis does not generate a verifiable prediction then it is useless.

In this section I shall present two conditions for a good test of a hypothesis and show that these take the sting out of Hempel's paradox. It should be understood that these conditions must be satisfied before the structure of the scientific program is relevant.

A good test of a research hypothesis must establish (1) that the prediction follows from the hypothesis, and (2) that if the hypothesis is not true, the prediction does not follow.

3.2.1. The First Condition

The conditional to be evaluated is

$$if \text{ [research HYPOTHESIS] } then \text{ [PREDICTION]}$$

or symbolically

$$if \text{ [H] } then \text{ [P].} \tag{3.1}$$

This is the first condition for a good test of the hypothesis. Recall that the conditional structure means that the consequent follows from the antecedent deductively. The strongest possible vehicle for this step is mathematics. Rigorous formal argument is rare in anthropology; typically recourse is had to verbal reasoning. This allows and seems to encourage misunderstanding both within and without the community of the discipline.

In testing, it is essential to realize that it is the hypothesis which is tested and not the prediction. Consequently the first condition for a good test is that the prediction follow in a deductive fashion from the hypothesis. This condition establishes that the hypothesis is necessary. In the event that the hypothesis is true and the prediction fails, the prediction is not a logical consequence of the hypothesis. One of the very common failings of anthropological research resides here. When the hypothesis is sufficiently vague that there may be some dispute about what it implies then it is unlikely that the prediction will follow deductively from the hypothesis. This matter needs considerable attention on the part of the investigator and failure to satisfy this condition will produce an inadequate test of the hypothesis.

3.2.2. The Second Condition

The second condition for a good test of a hypothesis is

$$if \ not \ [H] \ then \ \text{probably} \ not \ [P]. \tag{3.2}$$

This may be read as: if the hypothesis is not true then very likely the prediction is not true. This may seem a very strange condition until you realize the prediction is the only point of contact with the world. Note that the hypothesis cannot be justified by observing the prediction, for this results in the invalid argument structure affirming the consequent. Condition (3.2), then, provides the way out of this problem. It requires that the investigator be able to demonstrate that the prediction is extremely unlikely to occur if the hypothesis is not true. Then the second premise in the argument structure is [P]. If [P] is observed, that is the negation of *not* [P], then one validly infers the truth of the hypothesis. Failure to satisfy condition (3.2) of a good test of an hypothesis is a common logical violation perpetrated in the life and social sciences. Satisfying this condition establishes that the hypothesis is sufficient. A major segment of the research activity in the "soft" sciences is dedicated to condition (3.1). That is, the prediction is observed and from this it is then concluded that the hypothesis is true. You will recall that this is an invalid argument. The exclusive use of condition (3.1) is also responsible for the production of scientific politics. If, for example, there are two hypotheses, say, [H$_1$] and [H$_2$], both of which deductively support a single prediction [P] then how does one choose between them? Commonly the resort is to political activity with attendant appeals to authority and weight of evidence. (Politics also result from inadequate deduction of the prediction [P].)

   Clearly the prediction makes a statement about the world. A statement, recall, can be reliably determined to be either true or false. A prediction which cannot be checked using independent experimental or other means of observation is useless for the purpose of justifying an hypothesis. The hypothesis will be justified by the inductive scientific program. Recall that because this is an inductive argument it is possible that all the premises are true and the conclusion is false. Alternatively it is possible that the hypothesis is nowhere near being true and the prediction could come true by coincidence. This is a source of uncertainty which is due to induction. The other main source of uncertainty is due to the fact that the conclusion says the hypothesis is only approximately true. It is therefore possible that one could decide incorrectly about the truth of the hypothesis. (In Chapter 5 we shall have a great deal to say about this particular source of uncertainty.)

3.2.3. Failure to Satisfy Condition 2

3.2.3.1. HEMPEL'S PARADOX

The claim in Section 2.2.6 that any non-crow and non-black thing confirms that all crows are black is paradoxical only within the confines of deductive logic. Leaving aside the matter that *crow* is not a theoretical hypothesis, we may still see that neither of the two conditions is, or could be, satisfied. Specifically it cannot be shown that *if* [crow] *then* [black], nor especially can it be shown that *if not* [crow] *then not* [black]. So this paradox is simply a non-issue for science.

3.2.3.2. SMOKING AND LUNG CANCER

One of the most extensively studied phenomena in the modern world is

if [smoke cigarettes] *then* [probably lung cancer],

that is, smoking cigarettes increases the probability of lung cancer. Subject to accepted statistical criteria, this result has been repeatedly established. The scientific problems with this research are

(1) Most smokers do not get lung cancer, that is, it is questionable whether condition 1 has been met.
(2) Many victims of lung cancer never smoked or lived with someone who did; that is, condition 2 assuredly is not met.
(3) The argument structure is invalid—"affirm the consequent."

So unquestionably and without hesitation we may dismiss this as a scientific result.

3.2.3.3. EVOLUTIONARY THEORY

The inability of evolutionists to satisfy condition 2

if not [evolution] *then not* [man]

allows "Creation Science."

3.3. Examples

In the examples which follow, it was necessary to consider fields other than anthropology in order to present a complete sample. One need not focus on errors in deductive logic—blunders—required to produce the consequent from the antecedent in order to fill several volumes with examples of bad science. (Here it is worth recalling that the name of our discipline asserts that it is a science.) There are numerous examples of fallacious arguments as well. The reason for this wasteland is the incomprehensible poverty of theory in the discipline. There is no case of an anthropological theory which will support the process of producing a theoretical hypothesis. Without a hypothesis, there can be no verifiable prediction. Consequently it must be accepted that anthropology—at least—has not satisfied its obligation to its supporters to produce a science. This means that the consumers of anthropology—students and public—are not being provided with believable statements, and, so, must accept or reject statements on other than rational criteria. I suspect that in the absence of reason, one resorts to a variety of emotional crutches—such as the "weight" of evidence, or attributes of the proponent of the latest "theory."

By no means is anthropology unique in this regard. We are daily bombarded with solemn, authoritative, and contradictory, advice about nearly every aspect of modern living. The punch line for this process is "now you decide." This trivializes the matter—it becomes the equivalent of choosing a laundry detergent. It is also the case that myriad decisions must be made which cannot await scientific support.

3.3.1. Halley's Comet

Edmund Halley began applying Newtonian mechanics to the orbits of comets in 1695. His argument assumed that the sun was one focus of these elliptical orbits and that the bodies constituted a Newtonian system. With these concepts, he estimated that a comet he had observed in 1682 had a period of 75 years. In his search of the literature he found reports of observations of comets every 75 years back to 1305. In 1705 he predicted that the next appearance would be in December 1758. Halley died in 1743. The reappearance of the comet on Christmas Day, 1758 resulted in the general acceptance of the Newtonian tradition.

Initially Halley satisfied condition 1, that is, by assuming a Newtonian system he "predicted" past appearances of the comet. He established

if [Newton] *then* [comet has a 75 year orbit].

Then he addressed condition 2 by predicting the return of the comet in a specific 30-day period 53 years in advance. Probability theory was very primitive in the early 18th century, but we can see that, if the comet were returning randomly, the probability of a return within a specified 30-day period is about 0.002. Even in 1705 it was recognized that the predicted event was highly improbable unless the real system was behaving as a Newtonian system, that is

if not [Newton] *then* [highly improbable that the comet
returns in December 1758].

There was nothing, other than Newtonian mechanics, which could predict the time of arrival of comets. Since the predicted event occurred the following argument structure applies:

> *if not* [Newton] *then* [probably *not* prediction]
> [prediction]
> *thus* [Newton].

Note that verification was straightforward—either the comet appeared or it did not.

3.3.2. Mendelian Genetics

The experiments of Gregor Mendel constitute one of the turning points of biological science even though the report is highly questionable. Since all subsequent work has tended to support most of his results, it is not unreasonable to accept at least one of his "laws," e.g. the law of segregation. Two other "laws"—independent assortment and dominance—are demonstrably false in general.

At the time of Mendel's work, the accepted theory of inheritance was called "blending." This holds that offspring are a mixture, a blending, of the traits of both parents. (No one bothered about the fact that sex clearly is not blended.) This has a great deal of intuitive appeal since children do tend to resemble their parents. So Mendel's results were contrary to what "everyone knew" to be the case.

He investigated several different traits of garden peas, but in order to present the logic we need consider only one. Suppose a population of (sexually reproducing) organisms, some of which have the trait (TRAIT) and others do not (trait). Mendel created two different populations, both of which produced only one kind of offspring. Then he crossbred the two populations, producing a population of all TRAIT type. Next he interbred these plants and for every trait plant, three TRAIT plants were produced. The experiment and its results are

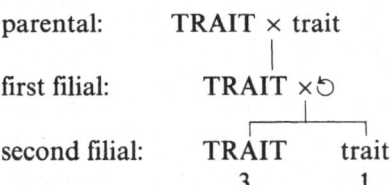

It is worth noting that Darwin was conducting a similar series of experiments at about the same time but he could not make sense of the second filial generation.

The parts of Mendel's report which are questionable are the counts of the second filial generation—the results are insufficiently variable. The direction of the "fudging" indicates that Mendel was committed to the 3 : 1 ratio of

TRAIT to trait before the experiment was begun. The reason Darwin was stumped was that he had no theory of heredity prior to his experiments. Mendel did. If one has no expectation, as Darwin did not, then a value of 3 TRAIT : 1 trait is no more or less meaningful than any other value. Mendel, on the other hand, had a very precise theory of heredity which specified that the only meaningful second filial generation result is 3 TRAIT : 1 trait.

Mendel's theory was: each parent (randomly) contributes exactly one of two possible "particles" to their offspring. It can be easily shown that, under the conditions of the experiment, the only allowable result is 3 TRAIT : 1 trait. Condition 1 is, then,

if [Mendel] *then* [3 : 1 in second filial]

and the structure of the argument is

if not [Mendel] *then* [*not* 3 : 1 in second filial]
[3 : 1 in second filial]
thus [Mendel].

Darwin's failure to obtain these results, at a time when he desperately needed to understand heredity, is a very poignant commentary on the price of "letting the data speak for themselves."

3.3.3. The Genetic Structure of Populations

In 1950 Muller defined what came to be called the classical model of population structure. Basically this model assumes that there is a "normal" gene for each genetic locus for each ecological niche. Among other things, this model predicts that most individuals in a given population are homozygous, they have two copies of the same "normal" gene at each genetic locus. Stable polymorphisms—more than one gene present in the population—should be rare. Genetic variation is a transient, if necessary evil.

By 1966 a technique called electrophoresis made possible the study of protein polymorphsim and monomorphism. This allowed Lewontin and Hubby to estimate the relative amount of polymorphism. The amount was found to be much greater than was predicted by the classical model.

Condition 1 is expressed as

if [classical] *then* [homozygous].

If a gene enhances the biological fitness of an organism, evolutionary theory predicts that it will increase until the entire population is homozygous. This prediction is obtained mathematically so we may accept that condition 1 is satisfied. The work of Lewontin and Hubby produces the following argument structure

if [classical] *then* [homozygous]
not [homozygous]
thus [*not* classical].

The classical hypothesis seems to have been rejected. The failure to satisfy condition 2, however, leaves the entire field of research in a most unsatisfactory state. The classical hypothesis does not uniquely produce homozygosity apparently.

Notice that the rejection of the classical hypothesis means that the research process must start from the beginning—a new hypothesis must be deduced, it must generate a prediction about the relative frequency of polymorphism, and then condition 2 must be satisfied. If sufficient effort had been devoted to condition 2, the rejection of the classical hypothesis would have been unnecessary. It would have been revealed to be inadequate. This would have produced a re-examination of the theory that produced the hypothesis.

3.3.4. Blood Pressure Change

A very common fallacy in anthropology is to suppose that because a hypothesis explains some event the occurrence of that event provides grounds for believing that the hypothesis is correct. You will note that this is simply the invalid argument structure called affirming the consequent and it is displayed below in the new notational structure:

$$\textit{if } [H] \textit{ then } [P]$$
$$[P]$$
$$\textit{thus } [H].$$

Examples of this particular fallacy are legion so only a single example will be given here.

In 1950 Hans Selye reported that perceptual stimuli of various kinds are capable of activating the release of certain hormones which are implicated in what he called the "General Adaptive Syndrome" of stress. This description of the reaction to stress on the part of the organism will be called a "Selye system." A great deal of research effort was directed subsequently to environmental stressors related to the clinical condition called hypertension. Alfred (1970) proposed that a change of cultural environment would activate the Selye system. He argued that if the Selye system is activated then both systolic and diastolic blood pressures should increase, though not necessarily enough to produce hypertension. Observations of blood pressure were taken on Navajo migrants in Denver, Colorado, and pre-migratory blood pressure readings were obtained from the health clinic on the Navajo reservation. The results were as expected. Both systolic and diastolic blood pressures increased between the two observational times. It was then concluded that migration was an activator of the Selyse system. The structure of the argument is

$$\textit{if } [\text{migration activates the Selye system}] \textit{ then}$$
$$[\text{blood pressure will rise}]$$
$$[\text{blood pressure rose}]$$
$$\textit{thus } [\text{migration activates the Selye system}].$$

This is an example of the invalid argument "affirm the consequent." This argument structure is not deductively valid and it is not a good inductive argument. In order that this particular piece of work could be salvaged, it would be necessary to develop the argument that if migration does not activate the Selye system then systolic blood pressure would not rise. In short, it would be necessary to show that the rise in systolic blood pressure would be an improbable event if migration had not activated the Selye system. Implicitly recognising the weakness of the basic argument, Alfred went on to comment on a variety of things, for example, no obvious changes in dietary behavior, water chemistry, altitude, or temperature. It was noted that sleeping patterns were probably quite different on the Reservation as compared to the urban pattern. This made it impossible to show that the observation was improbable if the hypothesis were not true. Notice that this particular project paid attention to a wide range of possible problems with regard to the conduct of research. Specifically, a great deal of time and effort was put into controls on the observations and to the techniques for determining whether or not the prediction had been observed. But none of that is sufficient to salvage the argument. Attention should have been given initially to condition (2) of a good test.

Also note that since Selye's pioneering work a huge literature has accumulated which generally supports the Selye system and the array of activators has been increased vastly. That is to say, the weight of evidence supporting the Selye system is great. However to my knowledge there has been no attempt to satisfy condition (2). Consequently one can not accept that the theory, the Selye system, has in fact been established.

3.3.5. Criminal Behavior Is a Mendelian System

Consider now an example which is logically more solid but is certain to be more controversial. The expectation that behavior, human or otherwise, has been subjected to evolutionary forces is in itself no longer controversial. The specific statement, however, that behavior is at least partially guided by genetics is highly contentious even though it is consistent with the expectations from evolutionary theory. The study of behavior genetics is a very active field today in the western world. The basic goal is to show that relatives behave more similarly to each other, independently of environment, than do nonrelatives. When it can be shown that this statement is (probably) the case then it is considered that a relatively strong argument has been presented for some kind of genetic involvement in the production of the behavior. Recently Mednick, Gabrielli, and Hutchings (1984) have considered the criminal behavior of adoptees in relation to the criminal behavior of both the biological as well as the adoptive parents. The components of this particular research effort are

if not [criminal behavior is not partially determined by a
  Mendelian system] *then not* [adoptive children will be
  more like their biological parents than their adoptive
  parents with regard to criminal behavior]
[adoptive children are more similar to their biological
  parents than they are to the adoptive parents]
thus [criminal behavior is partially determined by a
  Mendelian system].

It is important that you realise that behavior geneticists are not inherently dumb. That is, no one believes that evolution could have been acting so as to produce 20th century western criminality. They do, however, have the explicit expectation that evolution has been operating in a selective manner on some component of the genetics of *Homo sapiens* which by some unknown bio-chemical pathway leads to criminal behavior.

Also note that this particular test is an example of "deny the consequent." It is consequently a logically valid argument. Furthermore it is of great scientific as well as social importance. Notice that this project satisfied condition 2.

Most of the controversy surrounding modern work in behavior genetics centers on possible abuses of results of the work rather than the truth of the statements. It is certainly undeniable that results such as these can produce some very unacceptable social policy. The safeguards against abuse reside in the legal system, however. The possibility of abuse cannot be used to constrain inquiry.

3.3.6. Innate Principles of Geometry

Let's consider another example of a logically valid argument. Psychologists recently had the opportunity to evaluate the spatial orientation of a child who was blinded very shortly after birth (Landau, Gleitman, and Spelke, 1981). The child, Kelli, a two and one half year old girl, was asked to "go to" one of four different locations after having been familiarized with the experimental room. The components of this particular work are

if not [Kelli accesses innate principles of geometry]
    then [she will move about the room randomly]
  not [Kelli moved randomly]
  thus [Kelli is accessing innate principles of geometry].

The problem with this particular project is in the model of Kelli's behavior if she is not accessing these innate principles of geometry. The experimenters considered her, whatever her current location, to be at the centre of a circle. The circle was then arbitrarily sectored into nine arcs each subtending an angle of forty degrees radiating outward from Kelli. It was then argued that if Kelli is in fact choosing at random from among possible directions to move, the probability that she will choose the direction of the target is about one-ninth. This argument is unconvincing as, for example, momentum and/or current orientation would tend to impose a rather distinct bias on some of the directions. That is to say, one tends to have a preference for the direction in which one is moving or facing.

It will be noted that this particular example is one of "deny the consequent" and so is logically valid. In spite of its logical validity however we would be very hesitant to assert that the basic theory of innate Cartesian geometry has been established, or even strongly supported. The difficulty is that the prediction is not a logical consequence of the hypothesis, so neither condition 1 nor 2 has been satisfied.

3.3.7. Suicide as a Degenerative Disease

Few things have attracted the attention of romantic poets more than suicide. The basic assumption by them as well as subsequent social theorists, notably Durkheim, is that suicide is in some way a response to experience of the world. That is, life is considered to be abrasive and continually wears down the will to live. Ultimately one comes to a point at which the final restraint is eroded. This final event, whenever it can be observed, is often considered the precipitating event. If this is the case, then an appropriate model for the frequency of suicide is the standard epidemiological model of degenerative diseases. The distinctive feature of such diseases is that they are strongly dependent on age. That is to say, the frequency of the condition increases with age. Then if suicide is appropriately modelled this way we should expect to observe that the frequency of the event increases with age.

Recently we completed some work here on suicide. Specifically we studied the suicide behavior of American psychiatrists as reported in the obituaries published by the *Journal of the American Medical Association*. Several specialties were considered, but here I shall describe the results for psychiatrists only because of the notoriously high suicide rate in the profession. Also it was assumed that families of psychiatrists would be somewhat less inhibited than families of other medical practitioners in reporting a death as suicide as opposed to reporting it as an accident so there should be less error in the data. Note, however, that our concern was not to determine the rate of suicide. The goal was simply to determine whether the rate increases with age.

The components of this particular research effort are

> *if* [suicide is a degenerative disease] *then* [suicide rate
> will increase with exposure] (years of practice)
> *not* [suicide rate increases with exposure]
> *thus not* [suicide is a degenerative disease].

You will note that this is an example of "deny the consequent." It is therefore a logically valid argument structure. Inspection of the results indicate that the frequency of suicide actually decreases with exposure. This suggests quite a different model. It appears that the high rate in the early years of practice may be the result of the recruitment of suicide prone individuals into psychiatry. Note that "suicide-prone" may be interpreted as "partially genetically determined." Then at any given age the population includes survivors of normal mortality and the survivors of suicide. The frequency of suicides will decrease over time relative to normal deaths when: (1) the suicide rate is less than the normal mortality rate, and (2) the ratio of suicide rate to normal rate is approximately constant. These matters will be considered in detail later.

3.3.8. Kin Selection Theory and Mother's Brother

The fundamental assumption of kin selection theory is that the frequency of altruistic behavior is directly proportional to the coefficient of kinship. It is expected that siblings, for example, will be mutually altruistic about twice as often as cousins. Alexander (1979) has extended this argument to account for the common ethnographic observation of men being more altruistic to their sister's offspring than they are to the offspring of their wives. This phenomenon is to be expected when the confidence of paternity is less than or equal to 1/3.

The logic is as follows. Assume that the extramarital sexual activity is a constant such that all men are only $0 < p < 1$ confident that they are the father of their wife's offspring. In Figure 3.1, according to Alexander, the importance of mother's brother emerges when $f_{CE} < f_{CF}$ where f is the coefficient of kinship. If C were the father of E then $f_{CE} = 1/2$. But his uncertainty makes this value $f_{CE} = p/2$. His relationship to F is $f_{FC} = f_{CD}/2$. The relatedness of C and

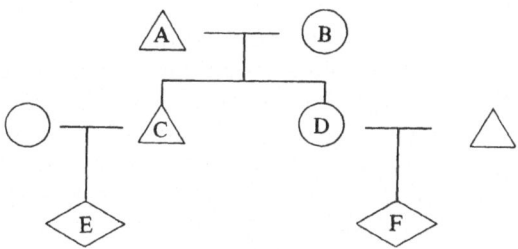

Overall coefficient of kinship
Probability of paternity = p
$f_{CE} = p/2$
$f_{CE} = f_{CD}/2 = (1 + p)/8$
$f_{CE} < f_{CF}$ when $p < 1/3$.

Figure 3.1

D is 1/4 from their mother, B, and p/4 from their mother's husband. Since the two lineages contribute additively $f_{CD} = 1/4 + p/4 = (1 + p)/4$. Then $f_{CF} = (1 + p)/8$. Now the event of interest is expected to occur when

$$p/2 \leq (1 + p)/8$$

or

$$p \leq 1/3.$$

Kurland (1979) and Gaulin and Schleglel (1980) have attempted cross cultural tests of this hypothesis. The latter used inheritance of property, real and movable, residence pattern, descent system, and succession to headman office. The results are as predicted. The crude level of measurement adds strength to the outcome. The argument is

> *if* [kin selection theory and confidence of paternity
>   threshold] *then* [reduced investment in wife's offspring]
> [reduced investment in wife's offspring]
> *thus* [kin selection theory]

which, however interesting, is invalid.

3.3.9. Fraternal Polyandry in Tibet

Polyandrous mating is rare among animals and especially so for humans. Murdock (1949) reports a frequency of less than 1% in the World Ethnographic Sample. When a rare structure is stable, i.e. not a transient event, it has great theoretical interest. Also the explanation of curiosities may overturn, or at least extend, current theory. Beall and Goldstein (1981) have issued a challenge to kin selection theory from the perspective of observations on fraternal polyandry in Tibet.

Polyandry appears to reduce the fitness of all of the males in the marriage and so would be expected only under conditions of resource deprivation. That is, when resources are so scarce that a given male experiences a fitness gain by joining his brother in a marriage over his expectation for his own monogamous marriage. Beall and Goldstein report (1) that polyandry is most common among the more affluent, (2) the average number of children ever born, and the number surviving, to monogamous females is greater than for polyandrous females. The authors conclude that "Tibetan fraternal polyandry does not appear to enhance the fitness of the individuals who practice it and in fact seems to entail substantial reproductive sacrifice" (*ibid*, 11).

The effort is flawed. Relatively minor is a typographical error in the equation for the probability of allele transfer—the "+" should be "−". More importantly is that, under the assumption of equal access by *n* brothers to a single female, the probability that a specific allele is transferred from a specific male is incorrect.

The fatal problem, however, is that the data refer to female reproductive success only. It seems clear that females do better in monogamous marriages than in polyandrous ones. If females were directing institutional formation, polyandry would have disappeared.

At several points the authors refer to tension developing within the household and brothers "fissioning" to form monogamous households. Studies of non-human primates indicate that a hierarchical structure will emerge within any multi-male unit. It is extremely unlikely that all brothers have equal access to the wife. If, as is more (not most) likely, younger brothers gain access proportional to their household rank, then it still seems that the top ranking male does surrender some reproductive potential. In doing this, however, it is by no means clear that his inclusive fitness is reduced as a relative (brother) is the beneficiary. In fact, from the perspective of a given brother, the production of four offspring by his siblings is equivalent to one of his own no matter which female is used. From the perspective of a younger brother, sexually but not culturally mature, any access to a female is a tremendous advantage. "Fission normally occurs when younger brothers first reach their early 20s, i.e. the normal marriage age..." (*ibid*). (The authors reject, without defence, and contrary to previous observations, the notion that polyandry is a temporary arrangement.)

The argument is

> *if* [kin selection theory and Tibetan polyandry] *then* [male
>   fitness enhanced]
> [female fitness is not enhanced]
> *thus not* [kin selection theory]

which is a non-sequitur.

3.4. Causation, Mill's Methods

In this section I wish to address the problem of causation. This is inherently a problem in inductive logic. John Stuart Mill was the first to systematize induction. As we will see shortly his techniques for recognizing causation are still very contemporary.

Correlation is a symmetrical relationship between two variables. For example consider two variables, say A and B, each of which has two possible values. The values of A are 'A' and '*not* A'; the values of B are 'B' and '*not* B'. 'B' is said to be correlated with A if they co-vary. A logical consequence of correlation is that the percentage of 'B' among 'A' is not equal to the percentage of 'B' among '*not* A'. Symmetry implies that if 'B' is correlated with 'A' then 'A' is correlated with 'B'. The converse is also true—that is, when 'A' is correlated with 'B' the percent of 'A' among 'B' is not equal to the percent of 'A' among '*not* B'. A very rough and ready kind of measure of the strength of the correlation is given by the difference in the percentages. (Notice that we will have ample opportunity in the sequel to formalize and extend these considerations.) The hazards of imputing causation to correlation are well known and there is a wealth of humorous and not so humorous examples. In cases where no embracing general theory exists, then there is no prior reason for assuming that either variable is a causal factor.

The observation that two variables tend to vary together may or may not be interesting but it certainly is not the end of the exercise. It is essential for the development of a mature science that causal factors be recognized and sequences developed. The main difference between causation and correlation is that the latter is symmetrical and the former is nonsymmetrical. That is, if 'A' causes 'B' then it is not the case that 'B' causes 'A'.

Mill presented three basic techniques for recognizing causation. These are called the method of agreement, the method of difference, and the method of concomitant variation.

The methods of agreement and disagreement are so commonly used together that it is natural to discuss them jointly. In the search for the cause of an event one typically looks at those cases where the event has been observed to occur as well as those cases where the event has been observed not to occur. The method of agreement is concerned with the first group, that is, those cases where the event is observed to have occured. Finding a common antecedent circumstance then is good evidence for its being a causal factor and particularly so when a general and embracing theory exists. This situation is

Case	Antecedent circumstance	Event
1	A, B	+
2	A, C	+
3	A, D	+
etc.	A,...	+

In every case the event is observed to have occurred and in every case of an occurence of the event the antecedent circumstance 'A' is similarly observed. 'A' is then a good candidate for a causal factor.

The method of differences requires simultaneous inspection of those cases in which the event is observed to have occured as well as those cases in which the event is observed not to have occurred. This situation is

Case	Antecedent circumstance	Event
1	A, B	+
2	B, C	−

In this display the event occurs when 'A' is an antecedent circumstance and does not occur when 'A' is not an antecedent circumstance.

The methods of agreement and difference are generalized in the method of concomitant variation. This method is taken by Giere as a definition of causation. Assume a population of possible observations which is randomly divided into two groups. Further assume that the cause is applied to all the members of one of the subsets and is not applied to any of the members of the second. If, then, the relative frequency of the event is different between these two subsets of the population good evidence has been obtained that the suspected cause of the event is in fact a causal factor.

Patrick Suppes develops the analysis of causation in the following manner.

(1) When the probability of the event 'A' given that the event 'B' has already occurred is greater than the probability of the event 'A' independently of the event 'B', then 'B' is said to be a *prima facia* cause of 'A'.
(2) When the probability of the event 'A' given that the events 'B' and 'C' have both occurred is equal to the probability of the event 'A' given that the event 'C' has occurred, then 'B' is said to be a spurious cause of 'A'.

A cause is said to be genuine when (1) it is a *prima facia* cause, and (2) it is not spurious. This latter condition is the devilish one. You should recognize this as the problem of establishing that the prediction, in this case the event 'A', is extremely unlikely if the hypothesis, the event 'B', is not true. Clearly one can never be certain that a factor, 'C', will not be discovered tomorrow or next week or the next decade which establishes that 'B' is simply a spurious causal agent in the production of the event 'A'.

3.5. Description, Pre-Science, Science

The conditions of a good test of an hypothesis produce a natural hierarchy of research.

1. Description—no hypothesis presented.
2. Condition 1 satisfied—pre-science.
3. Condition 2 satisfied—true science.

Description

All self-conscious disciplines probably advance along this hierarchy. At least it seems reasonable that they should. At the outset, before academic departments exist, people describe the world, they answer the question what is out there?

Condition 1 Satisfied

When sufficient information has accumulated, speculation about connections, between observations begins. This is the stage of explanation. The duration of this stage is probably dependent on the development of other "related" sciences. (The question of relatedness does not have an obvious answer. There is no way to predict the source of the next "breakthrough.") At some point in this exploratory stage, criteria for explanation become accepted and established. Anthropology is just entering the stage of establishing these criteria.

Condition 2 Satisfied

When the second stage is completed to the point that several, equally rigorous, distinctly different theories exist for explaining some event, the focus shifts to stage 3, showing that the event is uniquely explained by one of them. Each time this is done, conditions are ripe for a critical test, executing an experiment that exposes the theory to the most severe risk of falsification. When this process begins, the discipline is inexorably launched on the path of progress.

3.6. Summary

In this chapter we have presented some of the basic logical requirements of good research. This is a course about evaluating theoretical claims, that is to say, hypotheses. It is essential that one understand the distinction in the relationship between the theory and the evaulation of the theory, for this dynamic defines the scientific enterprise.

A concept which has not been addressed directly here but should receive some attention is that of reduction. It may be possible to develop a rigorous logic of science without the concept of reduction but the statements that result typically sound quite silly and contrived and are not easily tested. So we will adopt a weakly reductionistic approach to theory construction here. For us this simply means that causal factors, that is to say the hypothesis, are expressed in terms of units or forces which are in some sense "more fundamental" than the prediction. The prediction must make a statement about the world. The theoretical hypothesis then must be a description of the mechanism which produces the observed result.

Consider for example the hierarchical structure of phenomena which ranges from subatomic particles, atoms, molecules, cells, cell communities, organisms, ... It makes very little sense and is certainly quite uninteresting to attempt to explain the phenomena at one level in terms of other phenomena at the same level. One does not explain culture in cultural terms. The explanation of culture must be presented in terms of hominid evolution, especially of the brain. An attempt to explain patterns of property descent in terms of subsistence technology must then embrace the assumption that technology is in some sense more basic than the intergenerational transfer of wealth.

The production of statements about unobserved but observable phenomena necessarily means that they will sound somewhat "mystical" unless one is intimate with the logic. For example, in the 1930s Enrico Fermi included in an equation a term which was required by the assumption of basic symmetry in nature. Now in a real sense Fermi had no "right" to assume such symmetry. Today, this term has acquired a name—neutrino—and its very existence has only recently been confirmed. (Further, and incidentally, it is still not known if the particle has a rest mass.) Similarly, and again in the 1930s, Dirac predicted the existence, because it was logically required, of magnetic monopoles. This particle may have been observed once in 1981 and not since. The point of these examples is that the "outrageous" predictions produced by powerful theory do not lead to the collapse of civilization.

Now note that a testable theoretical hypothesis requires the prior existence of a theory. Without the theory there is no hypothesis. The hypothesis, when properly stated in dynamic terms, leads to a statement, a prediction, which can be tested by observing events in the world. Notice that the problem is not to establish the accuracy of the prediction but rather to determine the truth value of the theoretical hypothesis.

In much of this course we shall be concerned with statistical techniques which allow a decision about the truth value of a hypothesis to be made. An absolutely essential point which you must grasp is that the value of the statistic, in the sense of its worth or importance, resides entirely in the value (or worth, importance) of the theoretical hypothesis.

Theory without evaluation is totally fanciful, science fiction. It may be absolutely true but unless and until it produces procedures or observational techniques, and defines an experiment that exposes it to falsification, it cannot be seriously entertained as a theory. On the other hand, counting or observing without theory (empiricism) is a waste of resources. Note that theory and data are separate concepts but that neither can stand alone. That is, data are meaningful only in the context of theory, and theory is meaningful only in association with data. Often when a systemic structure such as this is specified there is a temptation to think of both (or all) parts as different manifestations of the same thing. This is not true, however. We shall maintain the distinction between theory and data while recognizing that both are essential to the scientific activity.

The standard technique of generating observations to be used in the statistical evaluation of a hypothesis is experimentation. In this milieu the factors of interest are known and are under direct control of the experimenter. It is rare, if it has ever happened, to be able to evaluate an anthropological hypothesis experimentally. So one assumes that "nature" is the experimenter and then by observing the "experimental" outcomes one attempts to infer what the factors are that are being varied. This activity is called sampling. In neither case are the causal factors known in advance except by hypothesis.

It is important that you realize that science is enquiry into the nature of the world. You may have heard that intuition (or some such) is equally valid. It is not because it is never tested directly. There are any number of apocryphal stories about individuals having seen the truth; and in the English-speaking world there is a distinct suspicion of rationality coupled with a strong preference for empiricism. Neither reason nor data is sufficient in and of itself. Both are required. "Theories" that are based on data only—a very common occurrence within anthropology—are *ad hoc*, they are tautologous (Section 3.3.2). Further, they are inherently weak because they are local either in time or space or both. A theory must make claim to universal validity or be trivial. The theory must also be stated in such a way as to allow in principle, refutation and it must also be repeatedly exposed to falsification.

Science gives no guarantee that the truth will ever be found. It does assume an objective real world that exists independently of our perceptions. With regard to ultimate truth and reality Einstein observed that "you can never open up the watch." That is, there is no way to determine what is true and what is false in any ultimate sense. At every decision point, then, we must always be prepared to be wrong. Only continuing experience can provide confidence in any given decision.

CHAPTER 4

Generating Predictions

4.1. Introduction and Orientation

Recall the first premise of condition 1 for a good test of a hypothesis:

if [H] *then* [P].

In this chapter we shall be concerned with the relationship between the theoretical hypothesis and the prediction. Specifically, when the theoretical hypothesis is stated mathematically then the power of centuries of true progress is available to the anthropologist for obtaining the prediction. Also recall the controversial assertion which was made earlier that in anthropology only stochastic theories are viable. In this section then we are concerned with modelling anthropological theories in the terms of probability theory. Later we shall have other recourse to probability theory for apparently different purposes but for now the goal is to model the theory.

Probability theory is a branch of mathematics (Feller, 1950) so it is concerned solely with relations among undefined things. To illustrate this concept Feller refers to the game of chess and notes that it is impossible to define the game other than by stating a set of rules. In this context it is meaningless to talk about the definition, or the true nature, of a pawn. Likewise geometry has no interest in what points or straight lines really are. Probability theory has been successfully applied to fields as diverse as physics and sociology, that is, from the sacred to the profane depending on your bias. For example, the behavior of particles immersed in a liquid is properly described by a probabalistic model called a random walk and named Brownian motion. Communications engineers can successfully predict the load on a telephone line at a given time of day. Ecologists can predict the composition of animal communities. Physical anthropologists can predict the genetic structure of populations. Sociologists can predict voting behavior. Psychologists can predict the waiting time to learning a particular behavior.

The reason that probability theory has been so successful in so many diverse fields is precisely because it is mathematical, i.e. its primary concepts are undefined. An undefined concept will be called a primitive. One such is a "random event." A random event is an empirical phenomenon characterised by the property that its repeated observation under a given set of circumstances does not always lead to the same observed outcome. Different outcomes result in such a way that there is a thing called statistical regularity. Probability theory then is the study of mathematical models of random events.

Anthropologists frequently find themselves in a position of dealing with variables which are so poorly defined that different observers of the same event can obtain different results. This should be a source of embarrassment. Things such as "a culture," a "descent pattern," etc. are often "discussed" with much more heat than light. Until this basic problem of making observations is settled the discipline will languish. However, here we need not concern ourselves with those difficulties. We have no substantive interest—neither theoretical nor empirical—so we can accept that the observations reported in the World Ethnographic Sample are, in truth, observations of the world.

Much of what follows will have the appearance of anthropology. Some of it actually is the real stuff. But while the topics and terms may be familiar and the results may seem interesting, important, trivial, or wrong, the sole purpose is to illustrate some elementary mathematical structures that inevitably will be increasingly useful as the field matures.

4.2. Background

4.2.1. The Sample Description Space

The sample description space of a random phenomenon is one of the primitive concepts in probability theory. One cannot give rules for the construction of a sample description space. Intuitively, however, the sample description space of a random phenomenom is the set of all possible outcomes of an experiment. (I will use the term experiment for the process of making an observation on a sample description space.) For the experiment "observe the sex of a birth" the sample description space is male or female. If the experiment is "observe the ABO blood group type of an individual" the sample description space consists of A, B, O, or AB. If the experiment is "observe a descent pattern in the World Ethnographic Sample" the sample description space is patrilineal, bilateral, matrilineal, or duolineal.

You will note that the number of possible outcomes of experiments on each of the sample description spaces described above is very small. Such spaces constitute the primary concepts of much of anthropology. The concept is not

restricted to small spaces, however, nor is it even necessary that they be finite. For example the experiment "observe an integer" can result in any of a
countably infinite number of possible outcomes. Nor is observability part of
the concept of sample description space. For example the experiment "count
the number of angels dancing on the head of a pin" is conceptually quite
sound, but only prior theory can specify whether the sample description space
is finite or infinite.

Dealing with such small sample description spaces means that we are able
to write down, using paper and pencil, all possible outcomes of an experiment.
Obviously this would not be the case for a sample description space infinite
in size.

The concept derives its primary importance from the fact that it provides
a means of defining an event. Whenever an experiment is made on a sample
description space, an event results. For the experiment "observe the sex of
a birth" either the event male or the event female will be observed. For the
experiment "observe a family form" from the World Ethnographic Sample
one of the following events may be observed: (1) independent families, (2)
minimal extended families, (3) small extended families, and (4) extended families.
Notice that with the logical relations of disjunction and conjunction it is
possible to define compound events on the sample description space. For
example we might choose to define a new event, small families, as the disjunction of independent families, minimal extended families, and small extended
families. Then we would have two events on the sample description space:
small families and extended families. The event small families would be observed
whenever at least one of the three defining family forms was observed. When
an event cannot be decomposed it is called a simple event; when an event is
composed of some relationship between several simple events it is called a
compound event. So now we have the definition of an event as a subset of the
sample description space. We can define for any event another event called
its complement. The complement of an event is everything in the sample
description space which is not the event. For example the complement of the
event small families is the event extended families. We might also define on
the sample description space for family form another compound event and
call it multi-generational. This event would be composed of: (1) minimal
extended families, (2) small extended families, and (3) extended families. If we
now form the disjunction of the two compound events small families and
multi-generational families we observe that the result is the sample description
space. Suppose that we form the intersection, conjunction, of these two compound events. That is, we ask for the event which is defined as small families
and multi-generational families. Notice that the conjunction of these two
compound events results in an event which is composed of the two simple
events, minimal extended families and small extended families.

Let us formalize these considerations a bit. We continue to use the sample
description space family form in the World Ethnographic Sample. Symbolize
the sample description space with S. Furthermore, assign the following labels

to the simple events of the sample description space:

Symbol	Simple event
A	Independent families
B	Minimal extended families
C	Small extended families
D	Extended families

The compound event small families now is seen to be defined as

$$E = [A \ or \ B \ or \ C]$$

and the compound event multi-generational families is defined as

$$F = [B \ or \ C \ or \ D].$$

Now note that the event defined by the disjunction of the two compound events E and F, [E *or* F], equals [A *or* B *or* C *or* D], that is, the sample description space. And the compound event defined by the intersection of E and F, [E *and* F], equals [B *or* C]. Note that each of the simple events of the sample description space is mutually exclusive. This means that all the intersections are empty. For example if we form the intersection of independent families and minimal extended families, that is, [A *and* B], we note that there are no events in the intersection. Alternatively we could attempt to form the intersection of the compound event E with the simple event D and again note that the intersection is empty. This constitutes a definition of mutual exclusiveness.

Now suppose that some numbers are associated with the simple events of the sample space. We impose the following conditions on these numbers:

(1) each number is positive and less than or equal to 1,
(2) the sum of all of the numbers is identically equal to 1.

When formalized these conditions are (they are called axioms) as follows:

1. $P(G) \geq 0$ for every event G
2. $P(S) = 1$ for the sample space.

In these axioms reference is made to a general sample space, not the specific one of family form. The event G should be considered any event; that is, it is not essential that G be a simple event. As innocent as these two axioms may appear you should be aware that they constitute the basis of probability theory. Probability theory is an investigation of the consequences of these axioms.

It is quite natural to assign probabilities to events in the sample space. Suppose, for example, the sample space given below:

$$S = \{a, b, c, d\}$$

and further suppose that we have defined on this sample space an event A which is composed of the simple events a, b, c, then the probability of A:

$$P(A) = P(a) + P(b) + P(c).$$

Notice that if each of the simple events has the same weight, that is, 1/4, then $P(A) = 3/4$.

There are several important things to notice about this particular definition. First of all it should be emphasized that one can only speak of the probability of an event when that event is defined as a subset of a specific sample description space. Also, while it is quite common to weight all the simple events in the sample description space equally this is by no means part of the definition.

A probability space will be defined as a sample description space on which a probability function has been specified. For the example above, the probability function assigned equal weights to all simple events in the sample description space. This definition of a probability space means that the probabilities may now be combined in a fashion which is strictly analogous to the rules for the construction of compound events from simple events. Specifically, assume the sample description space given above, that is the simple events $\{a, b, c, d\}$ on which the events $A = \{a, b, c\}$ and $B = \{c, d\}$ have been defined. Further assume that the probability function defined on the sample space is that of equal weights. Then

Sample description space	Probability space
$S = \{a, b, c, d\}$	$P(S) = 1.0$
$A = \{a, b, c\}$	$P(A) = 3/4$
$B = \{c, d\}$	$P(B) = 2/4$
not A $= \{d\}$	$P(not\ A) = 1/4$
A *or* B $= \{a, b, c. d\}$	$P(A\ or\ B) = 4/4$
A *and* B $= \{c\}$	$P(A\ and\ B) = 1/4$

4.2.2. Sampling

A conceptual device which is used by all probability theorists is the urn model. The main function of this particular model, which is nothing more than an urn with marbles in it, is to illuminate topics in sampling. This is effected by conceptually withdrawing marbles from the urn. We may assume that sampling may be done either with or without replacement with the obvious interpretations. If the sampling is assumed to be done without replacement, then the largest sample which can be drawn is equal to the number of marbles initially in the urn. On the other hand if after every draw the marble is replaced in the urn, then samples of any size may be obtained.

Suppose just for illustration we have an urn with three marbles in it: white (W), red (R), and black (B). We shall draw samples of size 2 from this urn. The possible outcomes of these two sampling strategies for this particular composition of the urn are

Without replacement
  (WB), (BW), (WR), (RW), (BR), (RB)
With replacement
  (WW), (RR), (BB), (WB), (BW), (WR), (RW), (BR), (RB).

Note that the difference between these two sampling strategies is that in sampling with replacement it is possible to get a sample of size 2 both of which are the same color. And the number of possible samples differs only by the number of different colored balls.

Lest you assume that this particular device, the urn model, is in some sense trivial you should be aware that Feller (1950) cites applications in fields as diverse as cosmic ray experiments, dice, coupon-collecting, aces in a bridge game, genetics, chemistry, irradiation and biology, and of course physics. Also you should not assume that a description of the sample description space or the probability space is a simple or straightforward matter. This will be discussed more fully below.

Now in order to motivate the analysis of the possible outcomes of sampling, let us assume that we shall be drawing samples of size n from an urn containing N distinguishable balls. The number of ways in which one can draw a sample of n balls without replacement from an urn containing N distinguishable balls is $N(N-1)\ldots(N-n+1)$. The number of possible samples available when sampling is done with replacement is N^n. Both these results are very easily demonstrated. Consider first the sampling regime without replacement. Then there are N possibilities for the first draw, $(N-1)$ possibilities for the second draw, and $(N-(n-1)) = N-n+1$ choices for the nth ball. When sampling with replacement the result is particularly simple: there are N possibilities on every sampling draw, so when a sample of size n is being drawn there are N^n possibilities.

Terms involving products of the sort $N(N-1)\ldots(N-n+1)$ are sufficiently common that it is useful to have notation for them. Following Parzen (1960) we shall use the notation $(N)_n$. Thus we define for positive integers N and n, when n is less than or equal to N:

$$(N)_n = N(N-1)\ldots(N-n+1).$$

We shall also have need of notation for the product of the first N integers. Following the general convention in probability theory, and elsewhere, we define the notation for the factorial, $N!$ as

$$N! = 1 \cdot 2 \cdot \ldots \cdot (N-1) \cdot N.$$

Now we can write $(N)_n$ in terms of factorial notation completely as

$$(N)_n = N!/(N-n)!$$

This definition holds for values of n from 0 up to and including N, which implies that $0! = 1$.

An important application of these results is that of finding the number of subsets of a given sample description space. You will note that numbers of the form $(N)_n$ are the number of ordered samples of size n that may be drawn without replacement. Using the example above, if we distinguish a sample (WB) from the sample (BW) then we are concerned with ordered samples, and terms of the sort $(N)_n$ apply. Usually however we are not concerned with ordered samples. Typically we are interested in samples with a specified number of, say, white marbles. Notice that this takes no account of the order in which they are obtained. Intuitively you might expect that the number of ordered samples will be greater than the number of unordered samples, that is, samples where order is irrelevant. And since we know how many ordered samples are present for a given sample size and total number of elements in the population, then knowing the number of ways that a particular sample composition can be ordered allows us to reduce that number to the number of samples where order is irrelevant. A moment's reflection should convince you that the number of distinct arrangements of n elements is exactly equal to $n!$. This may be seen as follows: Consider that we have three marbles, say one red (R), one white (W), and one black (B). How many ways can these three balls be sequenced? That is, if we record the first draw, the second draw, and the third draw, how many possible orderings of the colors is possible? You should see this as an example of sampling without replacement. This means that on the first draw there are three possibilities for the first color, on the second draw then there are two possibilities, and for the final draw there is only one possibility. Clearly for samples of size n this number of possible arrangements of the sample elements is exactly equal to $n!$ Now forming the expression

$$x_k = (N)_k/k!$$

we see that the number x_k is less than the number $(N)_k$ by exactly the factor $k!$. The numerator is the number of possible ordered samples of size k from the total number of N sample elements. And $k!$ is the number of possible arrangements available for k elements.

4.2.3. The Binomial Coefficient

We have just obtained a most important number called the binomial coefficient. The definition of this number in more common notational form is

$$B(N, k) = \frac{(N)_k}{k!} = \frac{N!}{k!(N - k)!} \tag{4.1}$$

This number appears repeatedly in what we shall be doing and so you should commit the definition to memory. Notice that it is simply the number of possible (unordered) subsets of size k that may be formed from the members of a set of size N.

A brief word about the notation used for the left-hand side of equation 4.1. Note the B indicates a particular function, a way of combining or relating some numbers. The components within the parentheses are the numbers to be related. These are called arguments. The function is defined by equation 4.1. It must always have exactly two arguments or it is undefined. When appropriate arguments are supplied, the operator, B, blinks once, coughs twice, whirrs thrice and delivers a number.

Let's consider a small example. We wish to form a committee of size three from a set of four available individuals. We identify the available individuals as $\{A, B, C, D\}$. Also note that when forming a committee we have no interest at all in the order in which sample elements are drawn. That is, as far as we are concerned the committee [ABC] is exactly the same as [CBA]. So, for this kind of problem, order is irrelevant. Now how many committees of size three can be formed? (In the binomial coefficient the rightmost argument, k, is called the "grab factor." This number clearly must always be less than or equal to the number on the left, N.) The set of distinguishable committees is

$$\{[ABC], [ABD], [ACD], [BCD]\}.$$

Notice that there are only four. Let us express this as

$$B(4, 3) = (4)_3/3! = (4 \cdot 3 \cdot 2)/(3 \cdot 2) = 4.$$

Now consider one of the possible committees, for example, the first, [ABC]. Notice that if we pay attention to order then there are six ways that this committee composition could have resulted. For example we could have drawn A first, B second, C third, or we could have drawn C first, B second, and A third, etc. You should convince yourself that there are six ways in which this committee could have resulted. This is true for each of the four committees listed above. Paying attention to order, there are $4 \times 6 = 24$ possible committees that could have been selected. But with regard to committees we consider that order of selection is irrelevant, so we reduce the 24 possibles by the number of ways of ordering each of the distinguishable committees.

You should think of this operation as that of partitioning the sample description space into two parts: one part which is committee, and a second part which is non-committee. Resorting once more to general notation, partition the sample description space, S, into a part with k members (which is committee), and the second part $(N - k)$ which is non-committee. It is very useful to be able to extend this concept into a partition of the sample description space into say r different groups. Let r be a positive integer and let k_1, $k_2, \ldots, k_r$ be positive integers such that $k_1 + k_2 + \cdots + k_r = N$. We shall now obtain a partition of the sample description space into r sub-samples such that the first has k_1 elements, the second k_2 elements, $\ldots$, and the last has k_r elements. Clearly r must be less than or equal to N. When r is equal to N, then each element of the sample description space is a sample.

Let's step this through. First we draw from the sample description space the elements of the first sample, that is, the first k_1 elements. The number of ways that we can obtain these k_1 elements is $B(N, k_1)$. There now remain $(N - k_1)$ elements in the sample description space available from which we will select the k_2 elements in the second partition. These elements may be obtained in $B((N - k_1), k_2)$ ways. Continuing in this fashion, the number of ways of obtaining the rth partition is $B((N - k_1 - k_2 - \cdots - k_{r-1}), k_r)$. The usefulness of the algebraic identity presented in definition (4.1) should now be readily apparent. Successive applications of this identity results in

$$
\begin{aligned}
B(N; k_1, k_2, &\ldots, k_r) \\
&= B(N, k_1) * B(N - k_1, k_2) * \cdots * B(N - k_1 - k_2 - k_{r-1}, k_r) \\
&= \frac{N!}{k_1! k_2! \ldots k_r!}.
\end{aligned}
\tag{4.2}
$$

You may satisfy yourself of the truth of this identity by noting that, for each pair of terms on the right of the equals sign, in the denominator of the leftmost of the pair is the term which is in the numerator of the rightmost of each pair. Terms of this sort will occupy much of our attention through this course. This is called a multinomial coefficient.

4.2.3.1. A SIMPLE COMBINATORIAL PROBLEM—THE "HYPERGEOMETRIC"

When the simple events of a sample description space are equally likely, a wide variety of problems may be solved using combinatorial methods. Consider the following problem: an urn contains six distinguishable (numbered) marbles, four of which are white and two are red. The white marbles are numbered 1 to 4, and the red ones are numbered 5 and 6. We shall draw samples of size two from this urn both with and without replacement. Let us find the following probabilities: (1) both marbles are white, (2) both are the same color, and (3) at least one is white. Define the event A to be both are white, the event B to be both are red, and the event C at least one of the marbles is white. So now the problem can be stated as finding (1) the probability of the event A, (2) the probability of the event A *or* B, and (3) the probability of the event C. Note that C = *not* B, so that $P(C) = 1 - P(B)$. Also note that the events A and B are mutually exclusive so that $P(A \text{ or } B) = P(A) + P(B)$.

First we find the total nunmber of points in the sample description space. (The notation $N(\cdot)$ will be used for the number of simple events in a particular event. You must be careful to distinguish this notation, where $N(\cdot)$ is a function, from N with or without subscripts, where the reference is to a specific number of things.) The number of events in each of the sample description spaces specified by sampling either with (R) or without (r) replacement is given as

$$\text{With replacement:} \quad N_R(S) = 6 \cdot 6 = 36$$

$$\text{Without replacement:} \quad N_r(S) = 6 \cdot 5 = 30.$$

And for reference we present below the simple event compositions of the two events A and B for sampling without replacement

$$A = \{(1,2),(1,3),(1,4),(2,1),(2,3),(2,4),(3,1),$$
$$(3,2),(3,4),(4,1),(4,2),(4,3)\}$$
$$B = \{(5,6),(6,5)\}.$$

You should convince yourself that when sampling with replacement, A would be as described above with the addition of $\{(1,1),(2,2),\ldots,(4,4)\}$, and the simple events of B would include $\{(5,5),(6,6)\}$.

Now let us determine the number of possible outcomes in the sample description space which satisfy the definitions of the two events. For small sample description spaces and small sample sizes, as in this example, it is feasible to list the possible outcomes satisfying the definition as above. We note there that for sampling without replacement, A is observed when any one of twelve pairs of simple events occurs. This is the point at which it is useful to see the application of the expressions given for sampling with and without replacement. The number of possible sample points in each of these sampling regimes is presented as

| | Without |
With replacement	replacement
$N_R(A) = 4 \cdot 4 = 16$	$N_r(A) = 4 \cdot 3 = 12$
$N_R(B) = 2 \cdot 2 = 4$	$N_r(B) = 2 \cdot 1 = 2$

Now it is possible to obtain the probabilities of each of the events as the ratio of the possible number of samples satisfying the definition of the event to the total number of points in the sample description space. Recall that sampling with replacement has a total of 36 (equally weighted) points in the sample description space, and sampling without replacement a total of 30 (equally weighted) points in the sample description space. In order to obtain the probability of A we form the expression $P_R(A) = N_R(A)/N_R(S)$. And for B we have $P_R(B) = N_R(S)$. The probabilities for each of the events are

Sampling with replacement	Sampling without replacement
$P_R(A) = 16/36 = 0.444$	$P_r(A) = 12/30 = 0.4$
$P_R(B) = 4/36 = 0.111$	$P_r(B) = 2/30 = 0.066$
$P_R(A \; or \; B) = P_R(A) + P_R(B) = 0.555$	
$P_r(A \; or \; B) = P_r(A) + P_r(B) = 0.466$	
$P_R(C) = 1 - P_R(B) = 0.889$	
$P_R(C) = 1 - P_r(B) = 0.934$	

You should note that the description of the sample description space which we have given in terms of marbles which are identifiable by color and number is not unique. For example, we could consider that the balls are distinguishable by color alone. In that case we get a sample description space, $S = \{(W, W), (W, R), (R, W), (R, R)\}$. In this case, assuming that all descriptions in S are equally likely then $P(A) = 1/4$, as does $P(B)$. Notice that the results obtained with this sample description space are radically different from those obtained earlier. You should be aware that the choice between these two sample description spaces and possibly others can only be made by (anthropological) theory. Both descriptions are equally valid mathematically. There is no prior reason to prefer one description over the other. Also note that the matter of choice of sample description spaces cannot be resolved by a simple appeal to "one works and the other doesn't." This is a meta-mathematical matter. In all of the rest of this course we shall assume that the first description of the sample description space is the proper one, that is, all marbles are distinguishable.

Let us consider that an urn contains N marbles of which N_w are white and N_r are red. Again we shall draw samples of size two with and without replacement. We ask for the following probabilities: (1) the first marble drawn will be white, (2) the second will be white, and (3) both will be white. Let us denote the event that the first marble drawn is white by A, that the second is white by B, and the event that both are white by the conjunction of A and B, that is, A *and* B. We further assume that the marbles in the urn are numbered from 1 to N with the white balls bearing the numbers 1 to N_w, and the red balls bearing numbers $(N_w + 1)$ to N.

When sampling with replacement the number of possible samples which can be obtained from this space is $N_R(S) = N^2$. And when sampling without replacement the number of points is $N_r(S) = N(N - 1)$.

Consider first sampling with replacement. The number of sample outcomes which satisfy the definition of the event A is $(N_w)(N)$ (because color on the second draw is irrelevant) and the number of possible outcomes satisfying the event B are $(N)(N_w)$. Since the two events referred to outcomes on the separate draws, the number of events in the conjunction of A and B is given by $N(A \text{ } and \text{ } B) = (N_w)(N_w)$. Now we may write the probabilities for these events as

$$P(A) = P(B) = N_w/N$$

$$P(A \text{ } and \text{ } B) = (N_w/N)^2.$$

Now consider the case of sampling without replacement. We note that now $N(A) = N_w(N - 1)$ since there are N_w possibilities for the first component in A and $(N - 1)$ possibilities for the second. Now we find the number of possible outcomes in the event B. Notice that it makes a difference if the first ball drawn was white or not. Assume that a white ball was obtained on the first draw. This could have occurred in N_w ways. Then for the second draw there are $(N_w - 1)$ possibilities. If the first marble drawn was red, which could have occurred in $(N - N_w)$ ways, then the number of ways of getting a white on the second draw is N_w. Since these two possible outcomes, white on the first draw and red on the first draw, cannot occur simultaneously, we form the number of possible sample outcomes in the event B by the disjunction, [(white on first *and* white on second) *or* (red on first *and* white on second)]. This is given by the expression $P(B) = N_w(N_w - 1) + (N - N_w)N_w = N_w(N - 1)$. Now we find the number of possible outcomes satisfying the event which is given by the conjunction, [A *and* B], of the two events which have been defined. The number of ways of getting a white marble on the first draw, thereby satisfying the condition for event A, is N_w. And then, given that a white was obtained on the first draw, the number of ways of obtaining a white on the second, and thereby satisfying event B, is $(N_w - 1)$. So the number of ways of simultaneously observing A and B is simply $N(A \text{ } and \text{ } B) = N_w(N_w - 1)$.

It is now a straightforward matter to obtain the probabilities of these events by forming the ratio of the number of ways of observing the event to the number of points in the sample description space.

4.2.3.2. AN ESP EXPERIMENT

Frequently we need to know how to tell a prediction from a guess. In the research related to the presence of extrasensory perception the following experiment is occasionally performed. Eight cards, four red and four black, are shuffled. The experimenter then looks at each successively. In another room, the subject attempts to guess whether the card the experimenter is looking at is red or black. The subject knows that the deck is evenly divided between red and black, and so will choose each color four times. Assume that the subject responded correctly on six of eight trials. We would like to know the probability of this outcome under the assumption that he has no extrasensory perception. We attend only those responses by the subject in which he made the response red. Then assuming that the cards are distinguishable and numbered from one to eight the subject's red responses can be written as the four card numbers (trials) which the subject said red. The number of points in the sample description space is given by $N(S) = B(8, 4) = 70$. In order that there be exactly six correct responses in the experiment, there must have been exactly three correct red responses. (Do you see why?) The subject responds red four times and black four times over the course of the experiment. If he achieves six correct from the eight trials, he must have made exactly one red error. Then the total number of ways of satisfying the condition of three correct red responses among four is given by the product $B(4, 3) \cdot B(4, 1)$. Notice that this is simply the number of ways of obtaining exactly three correct and one incorrect response out of the four. We may now find the probability of the event [six correct responses among eight] as the ratio of the number of ways of observing the event A to the number of points in the sample description space. This is presented as

$$P(A) = B(4, 3)B(4, 1)/B(8, 4)$$
$$= 16/70$$
$$= 0.23.$$

You will note that the probability of exactly six correct responses from among eight is very close to 1/4th which could be achieved by guessing alone. Consequently you would probably require more than six successes out of eight before being tempted to argue very forcibly for the existence of extrasensory perception.

Pierre Simon de Laplace (1749–1827) developed an entire theory of probability on this kind of analysis (Kac, 1964). In the form presented here it is called the hypergeometric distribution.

4.2.3.3. COMBINATORIAL MONKEYS

A primatologist has been on location long enough to recognize 4 animals confidently. During one day's observation he sees 1 of the 4 in a group of 5 animals. He wishes to know how many animals are in the area. Being a clever devil, he is willing to settle for a probability distribution. Specifically, he asks for the probability that there are $N = \{8, 10, 15, 20, 25\}$ resident monkeys given that he observed $K = 5$ animals of which $k = 1$ were familiar to him.

SOLUTION. He begins by reasoning that the 5 animals could have been sampled from the N in $B(N, 5)$ ways. Further, the observation of 1 of the 4 familiar animals could have been obtained in $B(4, 1)$ and the remainder in $B(N - 4, 4)$ ways. He then writes

$$P(N) = \frac{B(N - 4, 4)B(4, 1)}{B(N, 5)}$$

and proceeds to evaluate it for the values of interest.

N	$P(N)$
8	$((4!/4!0!)(4!/1!3!))/(8!/5!3!) = 0.0714$
10	$((6!/4!2!)(4!/1!3!))/(10!/5!5!) = 0.2381$
15	$((11!/4!7!)(4!/1!3!))/(15!/5!10!) = 0.4396$
20	$((16!/4!12!)(4!/1!3!))/(20!/5!15!) = 0.4696$
25	$((21!/4!17!)(4!/1!3!))/(25!/5!20!) = 0.4506$

Reviewing his results, he decides to accept, tentatively, the most probable value, i.e. $N = 20$.

4.2.4. Conditional Probability and Independence

Earlier when we were attempting to determine the probability of a red marble drawn on the second of two draws from an urn with red and white marbles the statement was made that "given that the first draw resulted in ...." In that circumstance we were attempting to determine the probability of one event after some prior event had already occurred. This is a conditional probability. Very frequently the knowledge that one event has occurred will modify our expectation for some subsequent event. This is obviously true when the early event is a causal factor producing an effect. The definition of conditional probability, however, depends in no way on the presence of a causal interpretation even though the concepts of independent and dependent events are defined in terms of conditional probability. These two concepts are of fundamental importance in allowing statistical decisions to be made.

Assume that two events, A and B, have been defined on a sample description space. Also assume that neither event is empty, that is, there are some sample points in both events. The event B is said to be statistically independent of the event A if the conditional probability of B *given* A is equal to the unconditional probability of B. Symbolically and for notation we introduce the following:

$$P(B|A) = P(B).$$

The vertical bar in the expression above is the notation that will be used for conditioning. This expression is a definition of independence. The event B is independent of the event A because its conditional probability is equal to its unconditional probability.

The definition of a conditional probability is

$$P(B|A) = \frac{P(A \ and \ B)}{P(A)}. \tag{4.3}$$

In words this may be read as the probability of the event B given that the event A has occurred is equal to the probability of the conjunction of the two events expressed as a ratio to the probability of the conditioning event. Using this definition we may express the probability of the conjunction as

$$P(A \ and \ B) = P(B|A) \cdot P(A).$$

When the two events are independent, then the conditional probability in the expression above may be replaced by P(B). This means that we have arrived at an expression for the probability of the conjunction of two events when they are independent:

$$P(A \ and \ B) = P(B) \cdot P(A).$$

In anticipation of what is to come, I should indicate to you that a major concern will be for the determination of independence and dependence between variables. Whenever the expression above is true the two events are independent. And whenever this expression is false the two events are said to be dependent, though the term nonindependent would be better (Parzen, 1960).

There is a common temptation to confuse independence with mutual exclusiveness. Consider for illustration the following example. An urn contains six marbles, four are white. Draw a sample of size two from the urn. Let A denote the event that exactly one is white and let B denote the event that both are white. Notice that the events are mutually exclusive, that is, there are no sample points which are simultaneously in both events A and B. We consider only the case of sampling with replacement because this is notationally most convenient. (You should convince yourself that the same result is obtained when sampling without replacement.) The sample space for this experiment is presented as

Sample:	(W W)	(W R)	(R W)	(R R)
Probability:	$(2/3)^2$	$(2/3)(1/3)$	$(1/3)(2/3)$	$(1/3)^2$

Notice that the unconditional probability of a white ball being drawn is $4/6 = 2/3$. The probability of the event A is just equal to the probability of getting a white on the first draw and a red on the second $(2/3)(1/3) = (2/9)$ or getting a white on the second and a red on the first $(2/9)$. This probability is $(4/9)$ since the two events are mutually exclusive. And the probability of the event B is simply equal to getting a white marble on both draws, $(2/3)^2$. This is equal to $(4/9)$ also. Since the two events are mutually exclusive, the probability of their conjunction must be identically equal to zero. You will note then that since neither of the events has probability zero it is impossible that the product of the probabilities can be equal to zero. It is therefore the case that events which are mutually exclusive are dependent (non-independent).

These concepts of independence and conditional probability can be extended to any number of events. Suppose, for illustration, a sample space with three events defined on it, $\{A, B, C\}$. The probability of C given that A and B have both occurred is defined as

$$P(C|A \text{ and } B) = P(A \text{ and } B \text{ and } C)/P(A \text{ and } B).$$

Notice that this definition holds only if $P(A \text{ and } B)$ does not equal zero. The events A, B, C are said to be independent if

(1) $P(A \text{ and } B) = P(A) \cdot P(B)$

$P(A \text{ and } C) = P(A) \cdot P(C)$

$P(B \text{ and } C) = P(B) \cdot P(C)$

(2) $P(A \text{ and } B \text{ and } C) = P(A) \cdot P(B) \cdot P(C).$

In order that three events be simultaneously independent it is necessary that they be pairwise independent as given in the first three equalities above, and that they be jointly independent as given in the last of the equalities above. Now, assuming that the probabilities of the events A and B and C, A and B, A and C, B and C, are all non-zero, independence implies that

$$P(A|B \text{ and } C) = P(A|B) = P(A|C) = P(A)$$

$$P(B|A \text{ and } C) = P(B|A) = P(B|C) = P(B)$$

$$P(C|A \text{ and } B) = P(C|A) = P(C|B) = P(C).$$

Notice that pairwise independence does not imply joint independence. The following example will illustrate this. Consider the sample description space $S = \{1, 2, 3, 4\}$. Define the events $A = \{1, 2\}$, $B = \{1, 3\}$, and $C = \{1, 4\}$. When all points in this sample description space are equally likely then $P(C) = P(C|A) = P(C|B) = 1/2$. But notice that $P(C|A$ *and* $B) = 1$. Since the conditional probability of C *given* (A *and* B) is not equal to the product of the probabilities of the simple events involved, the events are not jointly independent even though they are pairwise independent.

4.2.5. Distributions

A distribution is a functional rule that specifies the probability of each event in the sample description space. A functional rule may be defined in three ways:

1. An equation
2. A table
3. A graph

For many reasons, not the least of which is ease of manipulation, an equation is the preferred means of specifying a distribution. A table may be produced from an equation, and within computational limits, is a satisfactory representation. Typically an equation is not uniquely specified by a table, however, so analytic problems may be created in using this device. A graph adds two more sources of error—production (i.e. drawing) and interpretation (reading). There is no substitute for a picture when the purpose is to communicate qualitative features of a distribution, e.g. shape, dispersion, and location. But a picture cannot substitute for a table or an equation when the probability of an event is required. There is no substitute for a table in an environment without formidable computational support. But a table cannot substitute for an equation when the precision required exceeds that of the table, or when the probability of an event which does not appear in the table is required.

The notation of Section 4.2.3. can be used to motivate an important extension. Consider an urn that contains N distinguishable marbles, w of which are white and r are red. Let us determine the probability of exactly k white balls in a sample of size n obtained without replacement. Several features of the experiment are implied so far:

1. $N = w + r$
2. $n \leq N$
3. $0 \leq k \leq w$

A combinatorial structure can be used to obtain the desired probability:

$$P(k) = B(w, k)B(r, (n - k))/B(N, n). \tag{4.4}$$

Now we note that another condition obtains

4. $(n - k) \leq r$.

Note that the denominator of equation 4.4 is the number of ways of obtaining a sample of size n from among N; the left-hand term in the numerator is the number of ways of drawing k white marbles from the w in the urn; and the right-hand term in the numerator is the number of ways of getting the remainder of the sample. Inserting the definitions into 4.3 results in

$$P(k) = (w!/k!(w-k)! \cdot r!/(n-k)!(r-n-k)!)/(N!/n!(N-n)!).$$

Since $r = N - w$, the right-hand term in the numerator may be written as

$$(N-w)!/(n-k)!(N-w-n+k)!.$$

Now the right-hand side of the equation may be rearranged as

$$P(k) = \frac{\dfrac{n!}{k!(N-k)!} \cdot \dfrac{w!}{(w-k)!} \cdot \dfrac{(N-w)!}{(N-w-n-k)!}}{\dfrac{N!}{(N-n)!}}$$

$$= B(n,k)\frac{(w)_k}{(N)_n} \cdot \frac{(r)_{n-k}}{(N)_n}.$$

(4.5)

Feller (1950) shows that, as n gets large ("in the limit")

$$(w)_k/(N)_n = p^k$$

and

$$(r)_{n-k}/(N)_n = (1-p)^{n-k}$$

where p is the probability of a white marble. Then, for large n

$$P(k) = B(n,k)p^k(1-p)^{n-k}.$$

(4.6)

When experiments are performed in such a way that the outcome of any one experiment does not influence the outcome of any other experiment the observations are independent. This assumption will be always satisfied when sampling randomly with replacement. For the purpose of illustration consider an urn which contains three marbles labelled a, b, c. An experiment consists of sampling the urn three times with replacement. The sample space for this experiment is presented as

(aaa), (aab), (aac), (aba), (abb), (abc), (aca), (acb), (acc)

(baa), (bab), (bac), (bba), (bbb), (bbc), (bca), (bcb), (bcc)

(caa), (cab), (cac), (cba), (cbb), (cbc), (cca), (ccb), (ccc).

Notice that there are $3^3 = 27$ points in this sample description space. Also notice that while this particular illustration so far has assumed that each of the marbles is equally likely to be sampled, this is by no means essential. We could have, for example, that the probability of obtaining a marble with the label "a" on it is equal to P_a, and the probability of obtaining a ball with "b" on it is P_b, and similarly for a ball with a "c" on it, P_c. Then the probability of the first sample point, that is, (aaa) is simply equal to P_a^3. Likewise the probability of the sample point (bca) is equal to $P_b \cdot P_c \cdot P_a$. Then the probability of any sample point in the sample description space is given as the product of the probabilities of the components.

Now let us define a function on this sample description space, $f_a(x)$, whose value is equal to the number of times a marble with an "a" label on it is sampled. The range of values for $f_a(x)$ is $(0 \ldots 3)$ that is, we may obtain a sample point with no "a" in it or a sample point which is all "a" or any number in between. Denote the value of $f_a(x)$ as r. And define similar functions for the number of b's, (denote its value by the letter s), and a function for the number of c's (value denoted by t). (Convince yourself that $r + s + t = 3$.) Consider a special case, say $r = 1$, $s = 2$, and $t = 0$. We can obtain the probability of an event $r = 1$ *and* $s = 2$ *and* $t = 0$ by $P_a(1) \cdot P_b(2) \cdot P_c(0)$. Now locate all points in the sample description space which have exactly one "a" and exactly two "b." You will note that there are three such points: (abb), (bab), and (bba). Each of these points has the same probability and, since they are independent, the probability of this event is just 3 times the probability of one of the occurrences. Then the probability of their conjunction is simply three times the probability of that kind of a point.

Now assume that the urn is to be sampled n times. Then from results obtained earlier we see that there are

$$\binom{n}{r \, s \, t} = \frac{n!}{r!s!t!}$$

possible points satisfying the requirement that $N(a) = r$, $N(b) = s$, and $N(c) = t$. (Note that this is just equation 4.2.) We may obtain the probability of any one of these points as

$$(r)(P_a)(s)(P_b)(t)(P_c).$$

The conjunction of these conditions is given by the product of the probability of any one of the points times the number of ways the point may be realized. (You should note that $r + s + t = n$.) Clearly also the number of different types of marbles in the urn may be generalized to any number. Notice that there are four possible values for the function, $(0, 1, 2, 3)$. The basic problem is the following: assume that we were to repeat this sampling of marbles from this urn drawing three and then observing the value of $f_a(x)$ a large number of times. We ask then for the average, or expected value, of the function. Consider the outcome 3a. When this is the result $f_a(x) = 3$, and since there is only one such outcome in the sample description space we assign the probability 1/27 to this result. Now consider the outcome 2a. Notice that this result can be obtained in six different ways: $\{(aab), (aac), (aba), (aca), (baa), (caa)\}$. And we assign the probability of 6/27 to this possible function value. Now consider the outcome 1a. This can occur in twelve different ways, and so we assign the probability 12/27 to it. And finally the functional value 0a is observed in 8/27ths of the time. These considerations are arrayed as

$f_a(x)$	$P_a(x)$
3	1/27
2	6/27
1	12/27
0	8/27

If we weight each possible value of the function by the probability of observing it we obtain the expectation

$$E(f_a(x)) = (1/27) \cdot 3 + (6/27) \cdot 2 + (12/27) \cdot 1 + (8/27)0 = 1.$$

(The notation E is used for expected value.) Not surprisingly, it is expected that the average number of "a" that will be obtained in any sample of size 3 from the urn which has three marbles in it is equal to 1. This should be consistent with your intuition since we are drawing three balls from an urn where exactly one-third of them carry the "a" label. The generalized notation for this quantity is

$$E(x) = \sum_{i=1}^{i=n} x_i P(x_i). \tag{4.7}$$

The notation "capital sigma" ($\sum$) is used as a summation sign and in this particular case we indicate that the summation is to occur over all possible values of x. Then the things that are being summed are the products of the values of the function times the probability of that particular value.

Suppose that g and h are two functions defined on the sample description space. Then

$$E(g \cdot h) = E(g) \cdot E(h), \qquad \text{g and h independent}$$

$$E(g + h) = E(g) + E(h) \tag{4.8}$$

$$E(cg) = cE(g), \qquad c \text{ a constant.}$$

It would not be very smart to entertain the possibility that simply because the expected number of "a" in our sampling procedure is 1, exactly one will be observed in every sample. It would be useful to be able to describe quantitatively the expected scatter of the observed number in the sample around this expected value if the experiment is repeated a large number of times. This is typically measured by a quantity called the variance of the function. The definition of variance is

$$Var(f) = E(f - E(f))^2 \tag{4.9}$$

where f is some function defined on the sample description space. In words, this quantity is simply the expected value of the square of the deviations of the value of the function from its expected value.

Let us find the variance of the number of "a" in the sample of size 3 which we are dealing with. We already know that the expected value of this function is 1. And we have obtained the probabilities of each of the possible outcomes for the function f. These quantities produce

$$\text{Var}(f) = (1/27)(3 - 1)^2 + (6/27)(2 - 1)^2 + (12/27)(1 - 1)^2$$
$$+ (8/27)(0 - 1)^2 = 18/27.$$

(This quantity will be interpreted shortly.)

Let there be two independent functions, g and h, defined on the sample description space. The variance of the sum of these functions is the sum of the variances

$$\text{Var}(g + h) = \text{Var}(g) + \text{Var}(h) \qquad (4.10)$$

$$\left.\begin{aligned} &\text{Var}(c) = 0 \\ &\text{Var}(cf) = c^2 \,\text{Var}(f) \\ &\text{Var}(f + c) = \text{Var}(f). \end{aligned}\right\} \; c \text{ a constant} \qquad (4.11)$$

With a bit of manipulation, it can be shown that

$$\text{Var}(f) = E(f^2) - (E(f))^2 \qquad (4.12)$$

which is easier to produce with a calculator.

In much statistical work it is common to express an effect in terms of its observed deviation from the average effect. Let us consider this briefly. Define the standardized function by

$$f^* = (f - E(f))/\sqrt{\text{Var}(f)}. \qquad (4.13)$$

Notice that here the standardized function is expressed as the deviation of the function from its expected value, in the numerator, divided by the square root of the variance of the function. This denominator will become sensible shortly. For the moment let us investigate the properties of the function f^*:

$$E(f^*) = (E(f) - E(f))/b = 0$$
$$\text{Var}(f^*) = (1/b^2)\,\text{Var}(f) = 1 \qquad (4.14)$$

where $b^2 = \text{Var}(f)$. Note that this standardized function has some very nice properties. Specifically it has an expected value identically equal to zero and a variance identically equal to one. This facilitates the comparison of deviations from different sources.

The main use of the variance is in the determination of whether an observation is close to, or distant from, its expected value. When an observation is distant from its expected value, then under some conditions we may claim that an effect has been observed.

4.2.5.1. BINOMIAL DISTRIBUTION

A binomial variable is one which has two values. Often these are called "success" and "failure" but they may as well be "yes/no", "heads/tails", "0/1", or any other convenient label. Some examples are presented below. Equation 4.6 defines the binomial distribution. When the number of trials (n) and the probability of a success (p) are specified, the probability of exactly k successes may be calculated, or looked up in a table.

EXAMPLES.

1. The records of a particular hospital indicate that 50% of the births there are males. The last 5 births were female. What is the probability of this happening if there has been no change in the sex ratio at birth?

SOLUTION.

$$\text{Probability of male,} \quad p = 1/2$$

$$\text{Probability of female,} \quad q = 1/2$$

$$\text{Probability of 5 females} = B(5,0)(1/2)^0(1/2)^5$$

$$= 1 \cdot 1 \cdot 1/32$$

$$= 0.03.$$

This result would qualify as an unusual event given its low probability. We may ask, however, about the probability of all possible outcomes,

Number of males (k)	Number of females	$B(5,k)$	$(p^k)(q^{5-k})$	Probability
0	5	1	.03	0.03
1	4	5	.03	0.16
2	3	10	.03	0.31
3	2	10	.03	0.31
4	1	5	.03	0.16
5	0	1	.03	0.03
				1.00

Notice that the distribution is symmetrical. The probability of an outcome increases as the number of males increases from 0 to 2, and then decreases. This is an artifact of equal probability for male and female births.

Suppose that the probability of a male birth is 2/3 which makes the probability of a female birth 1/3. Now what does the distribution look like?

Number of males (k) ($p = 2/3$)	Number of females ($q = 1/3$)	B(5, k)	$(p^k)(q^{5-k})$	Probability
0	5	1	0.004	0.004
1	4	5	0.008	0.040
2	3	10	0.016	0.160
3	2	10	0.033	0.330
4	1	5	0.066	0.330
5	0	1	0.132	0.132
				0.996

Now note that the probability of 5 females in the last 5 births is much less probable, and that the distribution is no longer symmetrical.

In their paper on Tibetan fraternal polyandry, Beale and Goldstein (1981) are concerned with the probability of a specific gene being transmitted of offspring if there are n male siblings sharing a single female equally. Assume that all are full siblings. The probability that a given brother has the gene is $1/2$. Then the probability that exactly k of the brothers have the gene is given by B(n, k)$(1/2)^n$. The probability that the gene is transmitted by exactly 1 is just $1/2$ the probability that k brothers have it under the condition that all have equal sexual access.

2. An area of Ethiopia is populated by 2 different species of baboons. A primatologist observes 10 baboons in an area thought to be characterized by 25% species X and 75% species Y. Determine the probability of all possible species sample compositions.

SOLUTION.

Probability of species X, $p = 1/4$

Probability of species Y, $q = 3/4$

Probability of k species X = B($10, k$)$(1/4)^k(3/4)^{10-k}$.

Number of cases of species X	Probability
0	0.0563
1	0.1877
2	0.2816
3	0.2503
4	0.1460
5	0.0584
6	0.0162
7	0.0031
8	0.0004
9	0.0
10	0.0

If he observed 6 or more members of species X, he might be tempted to decide that there has been some change in local populations.

3. In the region of Ethiopia described in example 2, the primatologist is searching for the center of population for species X. He has decided on the following strategy: along an east-west route, he will obtain a sample of 10 animals every 5 kilometers. If he obtains 2 or fewer animals of species Y he will assume that the population center of species X has been located. Find the probability of his deciding that the center has been located.

SOLUTION. Note that in this problem the true value of p is unknown at each sampling site. The conditional probability of observing 2 or fewer members of species Y given various frequencies of species X is required.

(1) Probability of 2 or fewer members of species Y, $P(y \leq 2)$ is given by

$$P(Y \leq 2) = P(Y = 0) + P(Y = 1) + P(Y = 2)$$

$$= \sum_{i=0}^{i=2} B(10, i) p^{10-i} (1 - p)^i.$$

(2) P(X)   P(Y ≤ 2)

P(X)	P(Y ≤ 2)
0.10	0.0
0.25	0.0035
0.50	0.1719
0.75	0.9965
0.90	1.0

Note that when $P(X) \geq 0.75$, it is highly probable that he will decide that the center of population has been located.

4. The primatologist has decided to incorporate a technique called "focal animal sampling" which requires that a specific baboon be observed for a specified amount of time, usually one hour. At the moment he seeks Amber. There are 10 animals in the area. How many must be checked before finding Amber?

SOLUTION. As stated the question cannot be answered. We can, however, answer a closely related one—what is the probability that the next animal to be observed is Amber? Notice that this is equivalent to sampling with replacement, an urn containing 9 white balls and 1 red ball. Sampling with replacement corresponds to the period during which the primatologist cannot yet identify all individuals. When all are identifiable, he samples without replacement. The problem is to determine the probability that the red ball is drawn on the kth draw. This is equivalent to exactly $k - 1$ failures before the final success.

Number of draws required to get the red ball	Probability of k draws	
1	$1/10$	$=0.1$
2	$(9/10)(1/10)$	$=0.09$
3	$(9/10)^2(1/10)$	$=0.081$
4	$(9/10)^3(1/10)$	$=0.0729$
5	$(9/10)^4(1/10)$	$=0.0656$
6	$(9/10)^5(1/10)$	$=0.0591$
7	$(9/10)^6(1/10)$	$=0.0531$
8	$(9/10)^7(1/10)$	$=0.0478$
9	$(9/10)^8(1/10)$	$=0.0431$
10	$(9/10)^9(1/10)$	$=0.0387$
$\vdots$		

You should note that this could go on for a large number of draws—it is possible, sampling with replacement, that Amber will not be located within the lifetime of the primatologist.

5. Hausfater (1975) determined that the length of the menstrual cycle for female baboons was 32.5 days. The duration of the estrus period of the cycle is 7 days. These observations produce the estimated "probability of estrus" $p = 7/32.5 = 0.21$. Since the intensity of aggression in a troop depends on the number of estrus females present, it is useful to be able to predict the number. If the troop is observed for N days, what is the probability of exactly $k = 0, 1, 2, \ldots$ estrus females on any given day?

SOLUTION. If the problem is restated in terms of an urn model, its solution is straightforward. An urn contains n white balls. A trial consists of withdrawing each ball (sample without replacement), rolling a 5 sided die (the probability of any face is $1/5$), and if a 1 is obtained replace the white ball with a red one. After all n have been removed, count the number of red balls, replace the red with white balls, return all n to the urn, and repeat the process. We have

$$P(k) = B(n, k)(p^k)(q^{n-k}).$$

Assume that $n = 10$, $p = 1/5$ and for $k = 0, 1, 2, 3$ determine P(k)

k	$P(k)$	
0	$1 \cdot (1/5)^0 (4/5)^{10}$	$= 0.11$
1	$10 \cdot (1/5)^1 (4/5)^9$	$= 0.27$
2	$45 \cdot (1/5)^2 (4/5)^8$	$= 0.30$
3	$120 \cdot (1/5)^3 (4/5)^7$	$= 0.20$

Hausfater found that this model predicted the number of estrus females rather well. Be that as it may, you should be aware that there is very poor conceptual agreement between the model and the known dynamics of the system. So while it is possible to use the model to predict an empirical observation, understanding the process which produced the events has not been advanced. We shall return to this problem later.

6. Altmann and Altmann (1970) attempt to "account for the distribution of [baboon] group sizes within a population, and to account for the fact that mean group size varies greatly from one region to another" (204). Initially they used the BIDE model which predicts group size from the rates of birth (B), immigration (I), death (D), and emigration (E). It is argued that death and emigration rates—loss—depend on group size, as does birth rate; but immigration does not in any obvious way. (Later we shall consider a challenge to this last assumption.) [N.B. There seems to be a typographical error in the manuscript so the following discussion is intended to develop the spirit of their argument.]

SOLUTION. Let p be the probability that the group increases by 1 member—this is the sum of the birth and immigration rates; and let q be the probability that the group does not increase—let this be the probability of no increase plus the probability of a decrease by death or emigration. Then for a group currently of size n, the probability that at the next time period it is of size $n + k$ is

$$P(n + k) = B(n + k, k)p^k q^n.$$

Altmann and Altmann concluded that this model does not fit the observations of baboon group size.

Expected Values for the Binomial Distribution

Notice that, where S_n is the number of times a "success" is registered in n trials,

$$E(S_n) = np$$
$$\text{Var}(S_n) = np(1 - p),$$

$$(4.15)$$

the expected value and variance of a binomial random variable. We shall have need of these properties later when testing hypotheses statistically.

Also it is useful to have a bit more notation. For the binomial probability distribution, use the definition

$$b(k; n, p) = B(n, k)p^k q^{n-k}.$$

$$(4.16)$$

4.2.5.2. POISSON DISTRIBUTION (SIMEON D. POISSON, 1781–1840)

When a sample of observations on a binomial variable is observed, the number of successes is a random variable. That is, for a given number of observations, n, and a given probability of a success on each observation, p, the number of successes which occur may be any number from 0 to n. The expected number of successes is $m = np$. The expression 4.4 can now be written as

$$P(k) = B(n, k)(m/n)^k (1 - (m/n))^{n-k}.$$

It can be shown (Parzen, 1960; Ross, 1972) that, as n gets large and p small

$$(1 - (m/n))^n \sim e^{-m}$$

where $e \sim 2.7183$, the base of the natural logarithms. Since k is small relative to n

$$B(n, k) \sim 1$$

and

$$(m/n)^k \sim 0.$$

With these limiting values,

$$P(k) = e^{-m}(m^k/k!), \tag{4.17}$$

with $m > 0$, and $k = 0, 1, \ldots$. This is the Poisson distribution. It should be evident that an important application is as an approximation to the binomial—when n is large, the evaluation of $B(n, k)$ is difficult.

Olkin, Gleser, and Derman (1980) provide an interesting illustration of the approximation. For $n = 10$ and $p = 0.1$, $m = np = 1$ then

k	0	1	2	3	4	5	6	≥ 7
Binomial	0.349	0.387	0.194	0.057	0.011	0.002	0.000	0.000
Poisson	0.368	0.368	0.184	0.061	0.015	0.003	0.001	0.000

Note that even with n small the approximation is quite good. Letting n get large and p small (so that $m = 1$) with $n = 100$, $p = 0.01$ results in

k	0	1	2	3	4	5	6	≥ 7
Binomial	0.366	0.370	0.185	0.061	0.015	0.003	0.001	0.000
Poisson	0.368	0.368	0.184	0.061	0.015	0.003	0.001	0.000

You might expect (correctly) that with $n = 1000$, $p = 0.001$ the approximation is even better. "In general the Poisson distribution with parameter m provides a good approximation to the binomial with parameters n and $p = m/n$

in cases when n is large and p is small, and when $m = np$ is of moderate size (say $m \leq 20$)" (*ibid*, 191). ... the number of flaws in capacitors, ... the number of defects per linear unit of wire and of rope, ... the number of beetle larvae, ... the number of fish caught in a day, ... the number of photons reaching the retina, ... bacteria counts, ... the number of defective teeth [in humans] per individual, ... the number of victims suffering from various specific diseases, ... the number of labor strikes, ... the number of words misread in a text, ... the frequency of earthquakes, ... wrong telephone connections, ... radioactive emissions (*ibid*, 187). Clearly approximating the binomial is an important and quite useful application of the Poisson distribution. The range of applications is, however, much more extensive.

EXAMPLES.

1. The population density in a restricted area is m per unit area. At a university, for example, the density might be $m = 1/2$ per square meter. What is the probability that 2 sampled units are (1) both empty, (2) at least 1 is occupied?

SOLUTION. The probability that both sampled square meters are empty is

$$P(k = 0) = e^{-2(1/2)}\frac{(2(1/2))^0}{0!}$$

$$= 1/e$$

$$= 0.34.$$

The probability that at least 1 is occupied is

$$P(k \geq 1) = 1 - P(k = 0)$$

$$= 0.63.$$

2. A book of 473 pages contains 257 misprints, typographical errors. What is the probability that 10 randomly selected pages contain no errors?

SOLUTION. There are $257/473 = 0.53$ errors per page.
The Poisson parameter, the expected number of errors in 10 pages, is

$$m = (257/473) * 10 = 5.3.$$

The probability of 0 errors in 10 pages is

$$P(k = 0) = e^{-5.3}\frac{(5.3)^0}{0!}$$

$$= 0.005.$$

3. It should be noted that the Poisson parameter m must be estimated from observations. Parzen (1960) presents the following observations on vacancies in the U.S. Supreme Court.

Number of vacancies (k)	Number of years with k vacancies (N_k)
0	59
1	27
2	9
3	1
> 3	0

Define $t = k * N_k = 0 \cdot 59 + 1 \cdot 27 + 2 \cdot 9 + 3 \cdot 1 = 48$. Note that $N = \sum N_k = 96$. Then the parameter is estimated by

$$m = 4 * (t/N) = 2.0$$

vacancies per presidential term.

The probability that a president will make no appointments in a 4 year term to the Supreme Court is

$$P(k = 0) = e^{-4}(2^0/0!) = 0.14.$$

We may also ask whether these observations actually do follow the Poisson Law. In order to answer, a statistical test of the goodness of fit between the predicted and observed years with k vacancies is required. The details of the test will be presented later but an impression may be gained from the results below.

Number of vacancies (k)	Probability of k vacancies	Predicted years with k vacancies over 96 years	Observed years out of 96 with k vacancies
0	0.61	58.22	59
1	0.30	29.12	27
2	0.08	7.28	9
3	0.01	1.21	1
> 3	0.01	0.17	0

Note that there is very good agreement between predicted and observed.

4. In the Mahale Mountains west of Lake Tanganyika, Kawanaka (1982) observed 15 cases of attempted, suspected or successful predation by chimpanzees over 15 months. Should we arrive at the site to assist in the research, we may wonder about (1) the probability that we observe an episode tomorrow, (2) if an episode occurred earlier today, what is the probability of an episode tomorrow?, and (3) how long we should expect to wait before seeing 3 episodes?

SOLUTION.
(1) Convert the observations to a daily rate. There are about 453 days in 15 months so we set

$$p = 15/453 = 0.0331$$

per day. Assuming, questionably, that the episodes are independent—participation in or observation of an episode does not change the probability of predation for any of the animals involved—allows us to estimate the Poisson parameter as

$$m = np = 1 * p = p$$

since we are concerned only with a single day. Then the probability of no episode tomorrow is

$$P(0) = e^{-0.03} = 0.97$$

and probability of at least 1 episode is

$$P(k \geq 1) = 1 - P(0) = 0.0326.$$

(2) Given that an episode occurred earlier today, the probability of one tomorrow is 0.0326 because the events are independent.
(3) The expected waiting time to the occurrence of one episode is $(1/0.0326) =$ 30.2 days. Since the episodes are occurring randomly, the expected waiting time before 3 are observed is $3 * 30.2 = 90.6$ days.

5. In the Mt. Assirik area of Parc National des Niokola-Koba of south-eastern Senegal, Tutin et al. (1983) estimated the chimpanzee population to be 25–30 animals with a density of $0.09/km^2$. Over 44 months of observation between 1976 and 1979 there were 367 sightings of chimpanzees. This represents about 4000 observation hours and 284 contact hours. The estimated probability of a contact

$$p = 284/4000 = 0.0710$$

per hour. If the waiting time between contacts is a Poisson distributed random variable, then for any given 1000 hours of observation the following should obtain:

Number of contacts, k	Expected number of hours with k contacts
0	931.9
1	66.1
2	2.3
3	0.1
4	0.0

This has implications for staffing requirements.

If the animals are randomly distributed through the area—estimated to be about 42 km^2—then

Number of animals, k	Expected number of square kilometers with k animals
0	38.4
1	3.5
2	0.2
3	0.0

6. It may be suspected that the group size of social animals is a Poisson random variable. That is, animals are observed to be in the vicinity of each other but it is a purely random phenomenon. This implies that sociability is a chimera, an artifact of the random movement of animals.

In the study cited in example 5 the following observations were obtained:

k	Group size	Observed frequency	Expected frequency
1	6–10	54	82.59
2	11–15	23	21.50
3	16–20	9	3.73
4	>20	3	0.49

The expected values were obtained by using the parameter

$$m = \sum k_i O_i / \sum O_i = 0.5206,$$

where O is the observed frequency. Similar observations by MacDonald (1982) on Proboscis (*Nasalis larvatus*) in Brunei produced

k	Group size	Observed frequency	Expected frequency
0	1–5	12	5.82
1	6–10	13	11.64
2	11–15	6	11.64
3	16–20	3	7.76
4	21–25	2	3.88
5	26–30	3	1.55
6	31–35	1	0.52
7	36–40	2	0.15
8	>40	1	0.05

The parameter value is 2.00. As it turns out, neither of these sets is a Poisson random variable by the statistical criterion. But more importantly, even had the observations fit the model, there is no theory about social groups which satisfies the requirements of the Poisson distribution.

7. The hospital in Penticton, B.C., receives about 1.5 cases of snakebite each year. As of the end of July 1985 (a record year for heat in B.C.), it had treated 5 cases. Find the probability of this event if the parameter is 1.5.

SOLUTION.

Number of bites	Probability
0	0.2232
1	0.3347
2	0.2510
3	0.1255
4	0.0471
5	0.0141
6	0.0035

As the event of 5 cases in a season is quite unlikely—14 times in a thousand years—the hospital staff is justified in being alarmed. Note that should another case be reported before fall, there would undoubtedly arise rumors that the authorities were covering up a snake invasion to protect the tourist industry.

Expected Values for the Poisson Distribution

It probably comes as no surprize that the mean of a Poisson random variable is the parameter. One of the more interesting facts about this distribution is that the variance is also the parameter. That is

$$E(k) = m$$

$$Var(k) = m \tag{4.18}$$

where m is the parameter of the distribution.

4.3. Processes

A process is, for our purposes, a distribution in time. That is, if there is a mechanism pumping out 0's and 1's (failures and successes), it is a binomial process. (This is properly called a Bernoulli process.) A mechanism which produces 0's and 1's such that the waiting time between 1's is exponentially distributed, the process is Poisson. A mechanism which produces events in such a way that the probability distribution depends on the last event, the process is called Markovian.

Anthropology, as a self-conscious discipline, began as a concern for process, for evolution—cultural and biological. The goal was to explain variability through time. Notice that the description of the variability was not the objective; it was rather to understand the mechanism driving the process. So the study of process, evolution, is the intellectual foundation of anthropology.

There is a very large number of classifications of the models developed by mathematicians and probabalists. In this section are presented only two model structures—those that seem to me to have the broadest applicability. The purpose is to make available a few of the concepts from this very rich field. In order to keep the mathematics simple, some assumptions are required.

(1) Time is considered to be discrete, not continuous. This results in periods— which may be a second or a million years depending on the requirements of the theory. If, for example, the period is 1 hour, then all events in the interval (0800 to 0900] are all considered to be simultaneous. (The notation $(x$ to $y]$ means that the period begins after x and ends precisely at y. The interval is "open" on the left and "closed" on the right.)
(2) The process is "stationary." To fix the concept, consider the current controversy in evolutionary theory. One group argues that the process has occurred at a steady rate, i.e. it is stationary. Another group argues that there have been periods, e.g. the Eocene about 35 million years ago, during which the rate of change has been much more rapid than during other times, i.e. the process is non-stationary. I mention this pointedly lest you should decide that these assumptions are so restrictive as to make the models useless. Some very powerful theory has been modelled within these constraints.

In this section only stochastic—probabalistic—models are presented. Later other kinds of models shall receive brief attention.

A counting process models the number of events which have occurred up to time t.

You will note that $N(t)$, the number of events up to and including time t, does not decrease—it is monotonic—for a counting process. The conditions which must be satisfied are

(1) $N(0) = 0$
(2) $N(t)$ is an integer
(3) If $s < t$ then $N(s) \leq N(t)$
(4) If $s < t$ then $N(t) - N(s)$ is the number of events that have occurred in the interval $(s, t]$.

Examples of counting processes are

1. The number of mutations which occur in a given lineage.
2. The number of Folsom points recovered.
3. The number of matings between ego and mother's brother's offspring (cross-cousins).
4. The number of suicides.
5. The number of immigrants.

Non-counting processes are

1. The number of mutations if "forward" and "backward" mutations are considered.
2. The number of Folsom points available for study if some are placed in private collections.
3. Total current population including births and deaths.
4. Net migration, i.e. immigrants and emigrants.

The process is said to have independent increments if the numbers of events in different, non-overlapping, time intervals are independent. If, for example, there are two intervals defined by the points in minutes 12, 17, 22, then the number of events in $(17, 22]$ must be independent of the number of events in $(12, 17]$. With regard to the examples of counting processes above: (1) all available evidence indicates the randomness, independence, of mutations under constant levels of radiation; (2) the number of Folsom points recovered depends heavily on the amount of time trained hunters spend looking for them; (3) if mating is random then the number of matings between cross-cousins should depend only on the number of such potential mates; (4) suicide seems to have periods of greater and less popularity; and (5) immigration probably depends on perceived opportunity at the destination. In examples (2) through (5) it is possible to specify a set of conditions such that the increments of the process are independent. For example, for a constant number of searching hours in New Mexico, the process characterizing Folsom point recoveries may have independent increments. Immigration may have independent increments for constant perceived opportunity.

It is now possible to be a bit more precise about the concept of stationarity. A process has stationary increments if the number of events occuring in an interval depends only on the length of the interval. Let $s < t$ and x a constant; then a process is stationary when the number of events in the interval $(s, t]$ has the same distribution as the number of events in $(s + x, t + x]$.

4.3.1. The Poisson Process

The Poisson process is a counting process. It is defined as: The counting process $N(t)$ for $t \geq 0$ is a Poisson process with rate $m > 0$, if

(1) $N(0) = 0$.
(2) The process has independent increments.

(3) The number of events in any interval of length t is a Poisson random variable with parameter mt. That is, for $0 \le s < t$ and $n = 0, 1, \ldots$

$$P(N(t + s) - N(s) = n)$$

$$= e^{-mt}(mt)^n/n!$$

(4.19)

Note that condition (3) guarantees stationarity and that

$$E(N(t)) = mt,$$

the expected number of events in an interval depends only on the length of the interval, which indicates why m is called the rate of the process.

You should note that the theory being modelled should allow the determination of whether the process is expected to be Poisson.

Assume that the process is Poisson with rate (parameter) m. This means that for the time unit of the process, it is expected that, on the average, m events occur each unit of time. Suppose, to fix the concept, the events are episodes of predation by chimpanzees as in example 4 of Section 4.2.5.1. There it was estimated that these events occur at the rate of 0.0331 per day. At this rate, then, it is expected that there will be, on the average, $1/0.0331 = 30.2$ days between episodes. This is the mean, expected, waiting time. Clearly the mean waiting time and rate of occurrence are just the reciprocal of each other. (If a car is travelling at 50 mph, then the waiting time to travel 1 mile is $1/50 = 0.02$ hours or 1.2 minutes.) For the Poisson process the expected waiting time, the mean time, between events is $1/m$. For a sequence of events observed to occur at $(t_1, t_2, \ldots)$ the time between events is

$$T_1 = t_1 - 0$$

$$T_2 = t_2 - t_1$$

$$\vdots$$

$$T_n = t_n - t_{n-1}.$$

For the Poisson process

$$1/m = 1/n \sum_n T_n.$$

A quantity frequently of interest is the waiting time to the kth event. Notice that this simply k/m. A moment's reflection will convince you of this—if the mean waiting time to one event is $1/m$, then the mean waiting time to the kth event is $k(1/m)$. For example, if immigrants are arriving at a location according to a Poisson process with $m = 0.5$ per day, how long should we expect to wait before the 8th immigrant arrives? Note that the expected waiting time between events is $1/0.5 = 2$ days; then it is expected to be $8(1/0.5) = 16$ days before the 8th arrival.

EXAMPLE. Clark and Mangel (1984a,b) incorporate a Poisson model for the food search-encounter component of their analysis of foraging and flocking strategies. The focus of their effort is birds, but the results have clear substantive implications for both contemporary non-human Primates and the early stages of hominid evolution. This model will, therefore, receive attention here and in later sections.

Consider some resource—for example food, or females—which affects the fitness of an individual. For now, assume that the resource items have a constant, non-renewable value, V, to the fitness of the individual. (Clark and Mangel's analysis is in terms of the weight of the item which may be translated as proportional to fitness.) If the resource is "patchily" distributed and the searching animal has imperfect information, he may locate resource items according to a Poisson process with rate m. Once a resource item is located, its consumption or utilization requires some time, t. Let the unit of time be i, then the probability of locating a resource item within the time interval $(t, t + i]$ is mi and the expected time to encounter an item is $1/m$. Then the expected long-run contribution to individual fitness is

$$f_m = \frac{V}{t + 1/m} = mV/mt + 1. \tag{4.20}$$

This equation asserts that: (1) the fitness of an individual who searches at rate m, f_m, is equal to (2) the value of a single item, V, divided by the sum of (3) the consumption time, t, and (4) the searching time, $1/m$.

To pique your interest for things to come, Clark and Mangel are able to show, among many other things that if animals search together, the expected group size is greater than the optimum.

In the remainder of this section are presented some additional properties of the Poisson process.

1. Joint Poisson processes

Ross (1972) gives an example which will win you a beer nearly every time. Suppose customers arrive at a bank according to a Poisson process with rate $m = 1$ per hour. Half the customers are male, and males and females arrive independently of each other. You are told that 100 males showed up during a given 10 hour day. How many female customers arrive? Most people will say that 100 females should have arrived. But this response is based on specious reasoning. Since males and females arrive independently, and since the expected total is 10 for the day of whom 1/2 are female, the correct answer is 5.

This illustrates an important property about compound Poisson processes:

The joint distribution of independent Poisson processes is a Poisson process. If separate Poisson processes produce a Poisson process, then they are independent.

EXAMPLES.

1. If immigrants to Vancouver arrive according to a Poisson process at the rate of 100 per week and if 5% are from Germany, what is the probability of no German immigrant between 1 and 15 October (2 weeks)?

SOLUTION. Note that $m = 100$ immigrants per week; $p = 1/20$, the probability of an immigrant of German origin; and $t = 2$ weeks.

The Poisson parameter is

$$mpt = (1/20)(100)2 = 10,$$

then the probability of no arrivals from Germany during the 2 weeks is

$$e^{-10} = 0.00005.$$

You might be curious about this result if the rate were in terms of days instead of weeks. In this case $m = 100/7 = 14.29$ per day and the Poisson parameter is

$$mpt = (1/20)(14.9)\,14 = 10$$

producing the same result.

2. A family of Yukon Athapaskan speakers has a single hunter—Al. (The name has definitely lost something in translation.) During the fall, Al hunts, on the average, 3 days per week and never more than once per day. What is the probability that he does not hunt for 7 days?

SOLUTION.
(1) Binomial

$$P(\text{no hunt for 7 days}) = b(0; 7, 3/7) = 0.0199.$$

(2) Poisson

$$mt = (3/7)7 = 3$$
$$e^{-3} = 0.0498.$$

You should note the disparity between the binomial and the Poisson estimates.

3. In the example 2, Al has a kill rate of 2 deer for every 10 days he hunts. What is the probability that he brings a deer back today? What is the expected waiting time to the next deer returned?

SOLUTION.

$$P(\text{deer}) = P(\text{kill}) * P(\text{hunt})$$
$$= (1/5) * (3/7)$$
$$= 3/35$$
$$= 0.0857.$$

Waiting time to next deer

$$t = 1/0.0857 = 11.6667 \text{ days.}$$

4. In example 3, what is the probability of at least one deer within the next 3 days?

SOLUTION.

P(at least 1 deer) = 1 − P(no deer)

P(no deer) = P(no hunt in 3 days) + P(hunt *and* no kill)

P(no hunt in 3 days) = $e^{-3(3/7)}$ = 0.2765

The P(hunt *and* no kill) with $t = 3$ is expressed as

P(1 hunt *and* no kill) + P(2 hunts *and* no kill) + P(3 hunts *and* no kill)

$$= \sum_k P(k \text{ hunts } and \text{ no kill})$$

$$= \sum e^{-(9/7)}\frac{(9/7)^k}{k!}e^{-(3/5)}$$

$$= 0.1517(\sum {}^{(9/7)k}/k!)$$

$$= 0.1517((9/7) + (9/7)2/2 + (9/7)3/6)$$

$$= 0.1517(1.2857 + 0.8265 + 0.3542)$$

$$= 0.3742.$$

So P(no deer in 3 days)

$$= 0.2765 + 0.3742$$

$$= 0.6507$$

and the P(at least 1 deer in 3 days)

$$= 1 - 0.6507$$

$$= 0.3493.$$

You should notice that the probability of deer meat has two independent components—hunting and killing—each of which is modelled by a Poisson process.

4.3.1.1. THE TRANSITION TO BIG GAME HUNTING

In this section will be presented one of several extensions of the Clark and Mangel model. The resource item is food, the value is presumed to be in Kilocalories (Kcal), fitness is proportional to the value of food consumed, and individuals attempt to maximize their fitness. Food items—game animals—are encountered randomly according to a Poisson process, but the population of game is large so that there is no appreciable resource depletion due to hunting. Each human is assumed to have a capacity of C Kcal per meal in the sense that he ceases eating when this value is consumed. The resource animal is assumed to be of constant size $B = kC$, i.e. each animal represents k meals. To fix the concepts, assume a meal is 1000 Kcal. A deer may dress out at about 30 kg and each kilogram contains approximately 3800 Kcal. One deer = $(3 * 3800 \text{ Kcal})/1000 = 114$ meals. By contrast, a rabbit at 1 kg becomes only 3.8 meals.

Let t_1 denote the time required to prepare and consume one meal, and, as usual, m is the Poisson kill rate. Each individual hunts alone but stays in communication with the other hunters. When anyone makes a kill, all n hunters converge and eat until they are either satiated or no food remains. Satiation will occur first if $k > n$; when $n > k$ then none are satiated.

Define

$$a = \min(1, k/n),$$

that is, for any kill, if there are more meals than hunters then $a = 1$ whereas if there are more hunters than meals then $a = k/n$.

For n hunters the total hunting rate is mn and the expected waiting time to the next kill is $1/mn$. Each hunter consumes aC of the kill, that is, all share equally in the kill and each eats until he is either satiated or the food is exhausted. The time spent eating is at_1. The fitness of each of n hunters is

$$f_m(n) = \frac{af}{at_1 + 1/mn}$$
$$= \frac{C}{t_1 + 1/m(\min(k, n))}. \tag{4.21}$$

Any change in t_1, feeding time will affect fitness—for example, competitive interference would increase t_1 and decrease fitness. Similarly a decrease in the kill rate would impact adversely on fitness. These effects are modelled by the factor a. Notice that if $n > k$, more hunters than meals, the effect is to decrease feeding time—an individual will increase his rate of feeding in order to consume as much as possible.

Clark and Mangel define

1. the optimal group size as that which maximizes individual fitness, and
2. the equilibrium group size as that which maintains an average fitness which is not less than that of an individual hunting alone.

Given that the more hunters the more quickly resources will be located, as long as the group size is less than or equal to the equilibrium, an individual increases his own fitness through group membership. However, when the group is greater than or equal to optimal size, the average fitness of the group is degraded by his joining.

This model makes some predictions of great importance to anthropology. Assuming that self interest motivates individuals, the model predicts that the normal state is for groups to be larger than optimal! At equilibrium there is no advantage in joining a group, but up to this group size an individual will improve his lot by joining. It matters not at all to the individual deciding to join or not that the other members of the group will be less fit if he joins—so long as his fitness is enhanced he cannot but choose membership over isolation.

Another consequence of the model is that for groups at equilibrium, all would do better if the group divided into two halves. But no individual has any incentive to break away so these fission events are expected to be rare.

Note that, from equation 4.21, when food is abundant, i.e. m is large,

$$f_m(n) = C/t_1$$

because $1/m \to 0$, and group membership is disadvantageous because t_1 is increased through competitive interference. So grouping is not expected. Conversely grouping is expected when m is small. Equation 4.21 was evaluated over a range of prey sizes (0.1 to 100 kg), feeding rates (0.01 to 0.5 per hr), and kill rates, m (0.006 to 0.1 per hr) for an average hunter size (10 to 50 kg), a range which includes the size of the early Australopithecines. At about 43 Kcal/kg/day (Latham et al., 1970) a 30 kg hunter needs about 1290 Kcal/day, or an average of about 54 Kcal/hr. A "meal" was arbitrarily defined as 1/3 the daily requirements. A profound consequence of this model is that

(1) when the prey is small (0.1 kg) there is no optimal group size, i.e. individuals are better off alone
(2) for 1 kg prey there are two (only) optimal groups, 3 and 4 with equilibrium at 6 and 14 respectively
(3) for prey 10 kg or larger, a wide range of group sizes exist.

From (1) above comes the distinctive Hamadryas baboon (*Papio hamadryas*) troop. The animal feeds almost exclusively on small seeds. When the resource is abundant, isolated individuals constitute the optimal group. When the resource is scarce, all individuals do better by joining a group no matter what its size.

From (2) emerges something like the earliest hominid groups. For a moderately abundant resource of 1 kg—about the size of some modern rabbits—and average hunters less than or equal to 30 kg the optimal group is 3 or 4 depending on feeding rate. It is noteworthy that the most common equilibrium size is 6—that is, so long as the group contains 5 or fewer, an individual will prefer group membership to isolation. To put this result in context, however, the parametric "window" producing these groups is quite narrow, with the result that optimal groups are quite rare for small hunters and prey. One expects either isolated individuals or large hamadrayas type groups depending on the abundance of the resource.

In Table 4.1 will be found some representative predicted group sizes for hunters of size 10 kg or greater and prey size 10 kg or greater. Interestingly, this distribution of sizes is independent of increases in hunter size even though

Table 4.1. Representative Group Sizes by Kill Rate and Food Handling Time

Encounter rate (m)	0.1		0.05		0.025		0.012	
Handling time (t_1)	Opt.	Equil.	Opt.	Equil.	Opt.	Equil.	Opt.	Equil.
0.125	3	6	—	—	—	—	—	—
0.063	4	14	6	30	—	—	—	—
0.031	6	30	8	62	11	100	—	—
0.01	10	98	14	100	20	100	28	100

the larger individuals have greater caloric needs. It is also independent of further increases in prey size. This means, for example, that the cause of the observed evolutionary increase in hunter size, above a threshold; is not the effect of the behavior modelled here. It should also be noted that all the groups allow for the satisfaction of the total caloric requirement of the individuals. Some conditions—e.g. kill rate 0.1 and handling time 0.01—provide a much larger caloric supply than is required by an individual. These conditions should be thought of as the origin of food sharing. For example, a 20 kg hunter needs an average of about 36 kcal/hr. A kill rate of 0.1 and handling time of 0.01 will return about 143 Kcal/hr or about 4 times the requirement of the hunter. The surplus would be available for strategic distribution.

I have referred to feeding time and kill rate. More appropriately these terms should be handling time and encounter rate. Handling time is the average amount of time required for preparing and consuming one meal. For example, 4 men kill an animal that dresses out at 10 kg. The dressing (skinning, gutting, etc.) requires 60 minutes, and cooking 60 minutes. The product produces about 37000 Kcal. A meal for a 10 kg individual is about 143 Kcal, so there are about 260 meals. The handling time per meal is then $260/120 = 2.2$ minutes or 0.04 hours. Also using the term encounter rate instead of kill rate allows for scavenging and gathering rather than hunting exclusively.

In Table 4.1 you will note that both optimal and equilibrium group size increase as encounter rate decreases. Also both increase with decreasing handling time. It seems possible to exclude as unrealistic those conditions without a clear equilibrium size. The reason is that these are the maximum sizes that allow all group members to obtain at least as much of the resource as they could get acting in isolation. For the evolutionary stage under consideration, anthropological evidence does not support the existence of groups larger than 100. Then we are left with a rather small range of both handling and encounter times which produce acceptable group sizes.

4.3.2. Markov Chains

In this section is presented an introduction to a kind of non-independent stochastic process. Recall that the events of a Poisson process occur independently—the occurrence of one event has no influence on the occurrence of another. A Markov process—called a Markov chain when there is a finite number of kinds of events—is characterized by a one-step dependence. That is, the probability distribution for the occurrence of events depends on the last event to occur—and only on it. A Markov chain is said to have a one-step memory. The events of a Markov chain are called states and the occurrence of an event means the system is in that state.

When events are not independent they are said to be dependent. In the development of an anthropological theory, or for that matter any theory, typically a time dynamic is explicitly included. That is to say, the theory tells us the change to be expected in a particular system over a specified period of time. Since we are considering only stochastic theories, that is to say, probabalistic ones, there is no expectation of inevitability in the systemic change expected. The state of the system at one time determines the state of the system at a later or subsequent time only probabalistically.

The essential concept of Markovian dependence is that the state of the system at any given time depends, in a stochastic way, only on the state of the system at the immediately previous time. Note that this explicitly excludes all earlier time periods.

When the states of the system have been defined, the transition probabilities from state to state have been obtained, and the initial conditions of the system determined, the analytic structure is then called a Markov chain. The consequences of this particular structure have been analyzed at great length and applied with considerable success to a wide variety of different fields. For example, in physics the model describing the molecular diffusion of gases is a Markov chain. Markov chains have been applied to genetics, learning theory, and social mobility (Kemeny and Snell, 1960). Also, the growth of populations, the growth of populations subject to mutation, the theory of epidemics, to name only a few, have been successfully treated by the theory of Markov chains (Barucha-Reid, 1960).

No attempt will be made here to give even a summary analysis of a Markov chain. Suffice to note that a large number of statistical quantities of interest about the consequences of Markov chains are available. Given the wide success in applying the basic concept in diverse fields, it seems reasonable to anticipate that anthropologists will likewise discover the power. Rather than sketch in the analysis of a Markov chain, I shall present some of the results by way of example. You should consult one of the excellent introductory texts such as Kememy and Snell (1960) for the details.

The examples which will be discussed through this section are

(1) *Mate selection with population structure.* The Purums are a small hill tribe along the Indo-Burma border. The tribe is structured into five exogamous sibs—Marrim, Makan, Kheyang, Thao, and Parpa (White, 1963). Members of a given sib do not select mates randomly from the other sibs, i.e. members of the Parpa sib select mates from the Marrim sib, but not from the others. Figure 4.1 reproduces White's description of the mating structure of the tribe. Notice that choice depends on sib membership.

(2) *Learning theory.* A standard experiment in learning theory is the following. In the arms of a T-maze (Figure 4.2) place different stimuli. Introduce an experimental animal—mouse, gerbil, monkey, human—into the stem, and observe its choice between the two competing stimuli (Atkinson *et al.*, 1965; Beauchamp *et al.*, 1985). Over a series of trials using the same pair of stimuli an animal's choice during a specific trial depends on the choice it made on the previous trial.

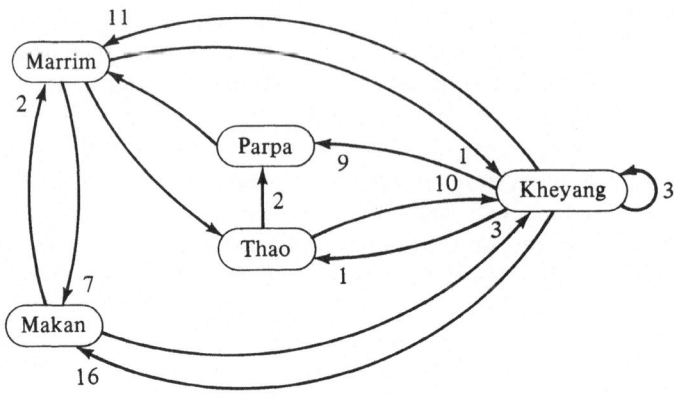

Figure 4.1

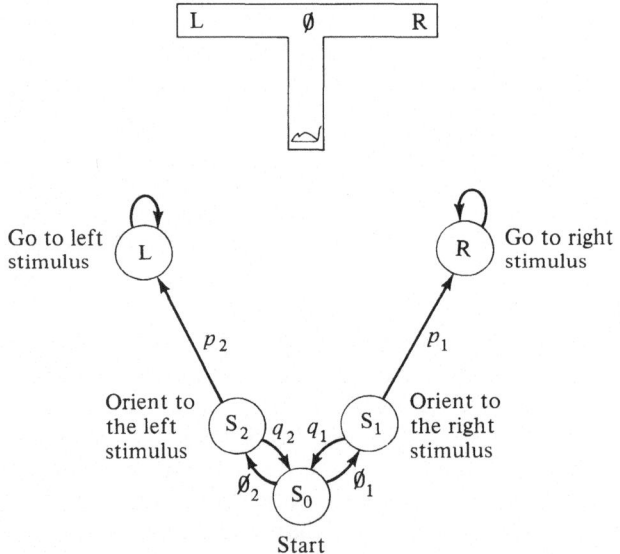

Figure 4.2

(3) *Modified Kula ring.* This version will use a system of four islands. Each island successively acts as host to a trading convention which is attened by trading partners from adjacent islands only. The conventions are held periodically at intervals of, say, one year. The object of each convention is to acquire ownership of a specific item, e.g. a necklace. The visitors bid for the item, and the current owner is not obliged to accept any offer. So the necklace may move to an adjacent island or remain in its current location.

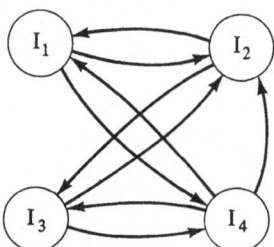

Figure 4.3

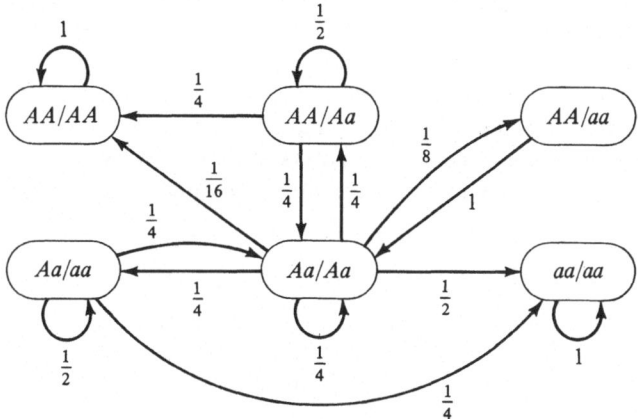

Figure 4.4

(4) *Mendelian genetics of sib mating.* This example is from Kemeny *et al.* (1959) and is useful for analyzing the production of a population of homozygotes. In this example each state is defined by the genotypes of a breeding pair. The locus under consideration has two alleles, A and a, on an autosomal chromosome. So the possible states are: AA/AA, AA/Aa, AA/aa, Aa/Aa, Aa/aa, aa/aa.

Introductory Concepts

The current state, or state of the system, refers to the last observed state. A Markov chain (finite number of states) may be visualized as a process which moves from state to state such that the probability of the next state depends only on the current state. Typically these probabilities are displayed in a transition matrix. Consider a system with 3 states: (S_1, S_2, S_3). Then the transition matrix is

$$
\begin{array}{cccc}
 & S_1 & S_2 & S_3 \\
S_1 & \begin{bmatrix} p_{11} & p_{12} & p_{13} \\ S_2 & p_{21} & p_{22} & p_{23} \\ S_3 & p_{31} & p_{32} & p_{33} \end{bmatrix}
\end{array}
$$

which has the following properties

(1) $0 \le p_{ij} \le 1$
(2) $\sum_j p_{ij} = 1$.

Note that the matrix has the same number of rows and columns, i.e. it is square. The rows represent the current state and the columns the next state. So p_{ij} is the probability of a transition from state S_i to state S_j. If a system is in state S_i then clearly $\sum_j p_{ij} = 1$. A matrix with this property is called stochastic. A Markov chain transition matrix is a square, stochastic matrix.

The matrix entries p_{ii} constitute the diagonal. If any of the p_{ii} is equal to 1 then when the system enters that state it is absorbed, i.e. there is no further change of state. The state i is called an absorbing state. If any of the p_{ij}, $i \ne j$ is equal to 1 then when the system enters state i it always goes to state j on the next transition. Then state i is called a reflecting state. For example

$$
\begin{array}{cc}
\textit{Absorbing} & \textit{Reflecting} \\
\begin{array}{cccc}
 & S_1 & S_2 & S_3 \\
S_1 & \begin{bmatrix} p_{11} & p_{12} & p_{13} \\ S_2 & 0 & 1 & 0 \\ S_3 & p_{31} & p_{32} & p_{33} \end{bmatrix}
\end{array} &
\begin{array}{cccc}
 & S_1 & S_2 & S_3 \\
S_1 & \begin{bmatrix} p_{11} & p_{12} & p_{13} \\ S_2 & 0 & 0 & 1 \\ S_3 & p_{31} & p_{32} & p_{33} \end{bmatrix}
\end{array}.
\end{array}
$$

First consider the absorbing matrix and assume that the system is currently in S_1. With probability p_{12} the system goes to S_2 and is absorbed. With probability p_{13} the system goes to S_3 and then, with probability p_{32} goes into S_2. With a reflecting state, with probability p_{12} the system changes from S_1 to S_2. Once in S_2 the system always goes to S_3 on the next transition.

A special case of a Markov chain is called a random walk. Consider

$$
\begin{array}{cccc}
 & S_1 & S_2 & S_3 \\
S_1 & \begin{bmatrix} 0 & 1 & 0 \\ S_2 & p_{21} & 0 & p_{23} \\ S_3 & 0 & 1 & 0 \end{bmatrix}
\end{array}.
$$

When in S_1 or S_3 the system always goes to S_2. Once in S_2 it goes to S_1 with probability p_{21} and to S_3 with probability p_{23}. There are many useful variations of this basic structure.

Bowers (Atkinson, 1965) used a random walk to describe the behavior of a rat in a T-maze learning experiment. "Cues in the T-maze can be classified into three sets: (1) those stimuli in the stem and vicinity of the choice point, denoted as the set S_0, (2) those stimuli likely to be sampled when the subject at the choice point orients to the right side of the maze, denoted as the set S_1, and (3) those likely to be sampled when the subject at the choice point orients to the left side, called the set S_2" (*ibid*, 155). When the choice is made between the experimental stimuli the experiment is terminated. Call these latter stimuli L and R. The experiment is diagrammed in Figure 4.2.

Notice that this experiment always starts in state S_0 and is absorbed in either state L or R. The transition matrix is

$$\begin{array}{c} \\ L \\ S_2 \\ S_0 \\ S_1 \\ R \end{array} \begin{array}{ccccc} L & S_2 & S_0 & S_1 & R \\ \left[\begin{array}{ccccc} 1 & 0 & 0 & 0 & 0 \\ p_2 & 0 & q_2 & 0 & 0 \\ 0 & O_2 & 0 & O_1 & 0 \\ 0 & 0 & q_1 & 0 & p_1 \\ 0 & 0 & 0 & 0 & 1 \end{array}\right] \end{array}.$$

This is a random walk because only transitions to adjacent states are allowed.

Two of the behaviors studied with this kind of structure are "vicarious trial and error," which are represented by sequences such as $(\ldots S_1 S_0 S_1 \ldots)$ or $(\ldots S_1 S_0 S_2 \ldots)$, and the number of transitions before being absorbed in either L or R.

If the matrix of transition probabilities is denoted $\mathbf{P}$ and the row vector Π_0 contains the initial probability of being in each state, then (Kemeny and Snell, 1960)

$$\Pi_n = \Pi_0 \cdot \mathbf{P}^n \tag{4.22}$$

which asserts that the key to the study of the changes in the vector Π is in the study of the powers of the matrix of transition probabilities.

In the remainder of this chapter some elementary matrix manipulation is required. All the necessary concepts are in Appendix A. Computation involving matrices is one of the greatest blessings computers can confer. Locating and learning one of many available, easy to use computer programs will be time well spent.

4.3.2.1. ABSORBING MARKOV CHAINS

If a Markov chain has at least one absorbing state and if it is possible to get to this state from all others (not necessarily in one step) it is an absorbing Markov chain. The transition matrices for the T-maze experiment and for the description of sib mating are examples of absorbing chains. When a process reaches an absorbing state it is said to be absorbed. It can be shown that the probability of absorption is 1.0, i.e. the process will terminate in finite time.

Some of the quantities of interest for such processes are

1. The probability that the process will terminate in a particular state.
2. The waiting time, i.e. number of steps, to absorption.
3. The number of steps the process is in a transient, i.e. non-absorbing, state.

All these quantities depend on the starting state.

The matrix of transition probabilities is reconfigured to have the following canonical (standard) structure:

$$P = \begin{bmatrix} I & 0 \\ R & Q \end{bmatrix}$$ (4.23)

where

I is the a by a set of absorbing states

Q is the t by t set of transient states

R is the t by a set of transient states

O is the a by t set of 0 probabilities.

Notice that

$$I = \begin{bmatrix} 1 & 0 & 0 \\ 0 & 1 & 0 \\ \vdots & & \\ 0 & 0 & 1 \end{bmatrix} \quad (a \text{ by } a)$$

which is called the identity matrix, and

$$O = \begin{bmatrix} 0 & 0 & 0 \\ 0 & 0 & 0 \\ \vdots & & \\ 0 & 0 & 0 \end{bmatrix} \quad (a \text{ by } t).$$

All absorbing Markov chain transition matrices can be structured this way. It can be shown that

$$P^n = \begin{bmatrix} I & 0 \\ * & Q^n \end{bmatrix}$$ (4.24)

where $*$ stands for the lower left hand (t by a) set which is not computed here. The entries of Q^n give the probabilities of being in each of the transient states after n steps for each possible transient state. As n gets large, the entries of Q^n get small, i.e. $Q^n \rightarrow 0$. This is to be expected since it was asserted earlier that the probability of absorption is 1.0.

The fundamental matrix is defined as

$$N = (I - Q)^{-1},$$ (4.25)

the inverse of the difference between the (t by t) identity matrix and the (t by t) matrix of transient state probabilities. The entries of N give the mean number of times the process is in each transient state for each possible transient starting state.

Consider the matrix of the T-maze experiment in canonical form

$$
\begin{array}{c}
\\
L \\
R \\
S_2 \\
S_0 \\
S_1
\end{array}
\begin{array}{ccccc}
L & R & S_2 & S_0 & S_1 \\
\end{array}
\left[
\begin{array}{ccccc}
1 & 0 & 0 & 0 & 0 \\
0 & 1 & 0 & 0 & 0 \\
p_2 & 0 & 0 & q_2 & 0 \\
0 & 0 & O_2 & 0 & O_1 \\
0 & p_1 & 0 & q_1 & 0
\end{array}
\right].
$$

Notice that the matrix of transient probabilities $\mathbf{Q}$ is the (3×3) lower right sector. Then

$$
\mathbf{I} - \mathbf{Q} =
\begin{bmatrix}
1 & -q_2 & 0 \\
-O_2 & 1 & -O_1 \\
0 & -q_1 & 1
\end{bmatrix}
$$

from which

$$
\mathbf{N} = (\mathbf{I} - \mathbf{Q})^{-1}
$$

$$
= 1/(1 - O_1 q_1 - O_2 q_2)
\begin{bmatrix}
1 - q_1 O_1 & q_2 & q_2 O_1 \\
O_2 & 1 & O_1 \\
O_2 q_1 & q_1 & 1 - O_2 q_2
\end{bmatrix}.
$$

Since this process always starts at the same point, S_0, the amount of time in the transient states is

Orient to left $S_2 : O_2/1 - O_1 q_1 - O_2 q_2$
Start $S_0 : 1/1 - O_1 q_1 - O_2 q_2$
Orient to right $S_1 : O_1/1 - O_1 q_1 - O_2 q_2.$

Now consider the second set of quantities mentioned at the beginning of this section, the number of steps in each transient state before absorption. Let $\mathbf{c}$ be a column vector of t 1's. In this case $t = 3$. Then the number of steps before absorption is $\mathbf{Nc}$.

In the example under consideration, since S_0 is always the starting state, the time before absorption is

$$
(1 + O_1 + O_2)/(1 - O_1 q_1 - O_2 q_2).
$$

And finally, the first set of quantities, the probability of absorption in a particular state is $\mathbf{NR}$. (Recall that $\mathbf{R}$ is the lower left-hand sector of the canonical structure.) Then for the example

Probability of absorption in the left arm $O_2 p_2/(1 - O_1 q_1 - O_2 q_2)$

Probability of absorption in the right arm $O_1 p_1/(1 - O_1 q_1 - O_2 q_2)$

where once more we take advantage of the constant starting state.

This model is very successful in predicting choice behavior in both paired, and larger, sets of alternatives. Further, the ratio of p_i/p_j—the ratio of preference of state i to state j—is strongly correlated with self reports of "confidence in the decision" among humans.

Now consider example 4, the genetics of sib mating. In canonical form the matrix of transition probabilities

$$\mathbf{P} = \begin{array}{c} \\ AA/AA \\ aa/aa \\ AA/Aa \\ AA/aa \\ Aa/Aa \\ Aa/aa \end{array} \begin{array}{cccccc} AA/AA & aa/aa & AA/Aa & AA/aa & Aa/Aa & Aa/aa \\ \left[\begin{array}{cccccc} 1 & 0 & 0 & 0 & 0 & 0 \\ 0 & 1 & 0 & 0 & 0 & 0 \\ 1/4 & 0 & 1/2 & 0 & 1/4 & 0 \\ 0 & 0 & 0 & 0 & 1 & 0 \\ 1/16 & 1/16 & 1/4 & 1/8 & 1/4 & 1/4 \\ 0 & 1/4 & 0 & 0 & 1/4 & 1/2 \end{array}\right] \end{array}.$$

The interpretation of the transitions is somewhat novel since each state represents two animals. The first two rows are obvious, so attend the third. This is interpreted as follows:

The mating of AA with Aa produces offspring in the proportions $(1/2)AA$, and $(1/2)Aa$, so for two offspring (the columns of the matrix) the proportions are $1/4 = (1/2)^2 AA/AA$, $1/2 = 2(1/2)^2 AA/Aa$, $1/4 = (1/2)^2 Aa/Aa$. Notice that the mating produces no aa offspring so the pairs involving this genotype necessarily have probability 0.

Also note that the mating AA/aa produces only Aa offspring. Interpretation of other entries proceeds similarly.

Now we must find $\mathbf{N}$, the fundamental matrix. First form the difference

$$\mathbf{I} - \mathbf{Q} = \begin{bmatrix} 1/2 & 0 & -1/4 & 0 \\ 0 & 1 & -1 & 0 \\ -1/4 & -1/8 & 3/4 & -1/4 \\ 0 & 0 & -1/4 & 1/2 \end{bmatrix}$$

and then

$$\mathbf{N} = (\mathbf{I} - \mathbf{Q})^{-1} = \begin{bmatrix} 2.67 & 0.17 & 1.33 & 0.67 \\ 1.33 & 1.33 & 2.67 & 1.33 \\ 1.33 & 0.33 & 2.67 & 1.33 \\ 0.67 & 0.17 & 1.33 & 2.67 \end{bmatrix}.$$

If, for example, the initial breeding pair is AA/Aa, then before absorption, a pair of offspring consisting of an AA and an Aa is expected to occur 2.67 times from the mating; likewise, a pair AA/aa is expected only 0.17 times.

Now we check for the number of steps expected before absorption. Notice that absorption means that only homozygotes are being produced:

$$\mathbf{Nc} = \begin{array}{c} AA/Aa \\ AA/aa \\ Aa/Aa \\ Aa/aa \end{array} \left[\begin{array}{c} 4.83 \\ 6.67 \\ 5.67 \\ 4.83 \end{array} \right].$$

If the initial pair is AA/Aa or Aa/aa then, on the average only 4.83 matings are required before absorption, 5.67 are expected if the initial pair is Aa/Aa, and 6.67 if the initial pair is AA/aa.

And finally we check for the probability of being absorbed in each of the absorbing states:

$$\mathbf{NR} = \begin{array}{c} \\ AA/Aa \\ AA/aa \\ Aa/Aa \\ Aa/aa \end{array} \begin{array}{cc} AA/AA & aa/aa \\ \left[\begin{array}{cc} 0.75 & 0.25 \\ 0.50 & 0.50 \\ 0.50 & 0.50 \\ 0.25 & 0.75 \end{array} \right]. \end{array}$$

If the initial pair is AA/Aa the probability that the ultimate state is the production of AA homozygotes is 0.75 and only 0.25 for the aa homozygotes. Notice that if the initial pair is AA/aa or Aa/Aa the system ultimately produces either AA or aa homozygotes with equal probability.

4.3.2.2. Non-Absorbing Markov Chains

A state j is accessible from state i if it is possible to get from i to j in a finite number of steps. If i is an absorbing state, then it is not possible to get to any other state once i is entered. Without absorbing states, then it is possible to get to any state.

Consider a transition matrix $\mathbf{P}^{(1)}$. The superscript identifies the number of steps represented by the matrix. It is easy to show that

$$\mathbf{P}^{(2)} = \mathbf{P}^{(1)} \cdot \mathbf{P}^{(1)} = \mathbf{P}^2.$$

It can also be shown that the n-step transition matrix is

$$\mathbf{P}^{(n)} = \mathbf{P}^n.$$

When $\mathbf{P}^{(1)}$ has no absorbing states then there will be no zeros in $\mathbf{P}^{(n)}$ for some value of n; thus it is possible to get from any state to any state eventually. When states are mutually accessible they are said to communicate. A most remarkable result is that as n gets large ("in the limit"), each row of the matrix becomes the same vector with all entries greater than 0. Let $\mathbf{P}_{ij}^{(n)}$ be an entry in the n-step transition matrix—the probability that the system is in state j after n steps if it started in state i. Then

$$\Pi_j = \mathbf{P}_{ij}^{(n)}, \qquad n \to \infty.$$

The long-range predicted state does not depend on the starting state.

It is useful to be able to determine the entries of the n-step transition matrix. Consider the hypothetical transition matrix for the passage of a necklace through the modified Kula ring. The islands are identified as I_i:

$$
\mathbf{P} = \begin{array}{c} \\ I_1 \\ I_2 \\ I_3 \\ I_4 \end{array}
\begin{array}{cccc}
I_1 & I_2 & I_3 & I_4 \\
\end{array}
\left[\begin{array}{cccc}
0 & 3/8 & 0 & 5/8 \\
4/8 & 0 & 4/8 & 0 \\
0 & 1/8 & 0 & 7/8 \\
3/8 & 3/8 & 2/8 & 0
\end{array} \right].
$$

Recall that trading conventions are held annually and are hosted by the current owner. Conventions are attended only by trading partners from the adjacent islands. Note that the trading competitors are not equal, e.g. when Island 3 is the owner, the partner from Island 4 acquires it 7/8 of the time. We shall determine the probability distribution for the first few trading episodes:

$$
\mathbf{P^1} = \left[\begin{array}{cccc}
0 & 0.3750 & 0 & 0.6250 \\
0.5000 & 0 & 0.5000 & 0 \\
0 & 0.1250 & 0 & 0.8750 \\
0.3750 & 0.3750 & 0.2500 & 0
\end{array} \right]
$$

$$
\mathbf{P^2} = \left[\begin{array}{cccc}
0.42 & 0.23 & 0.34 & 0.0 \\
0.0 & 0.25 & 0.0 & 0.75 \\
0.39 & 0.33 & 0.28 & 0.0 \\
0.19 & 0.17 & 0.19 & 0.45
\end{array} \right]
$$

$$
\mathbf{P^3} = \left[\begin{array}{cccc}
0.12 & 0.20 & 0.12 & 0.56 \\
0.41 & 0.28 & 0.31 & 0.0 \\
0.16 & 0.18 & 0.16 & 0.49
\end{array} \right]
$$

$$
\mathbf{P^6} = \begin{array}{cccc}
0.26 & 0.26 & 0.20 & 0.28 \\
\end{array}
\left[\begin{array}{cccc}
0.26 & 0.25 & 0.21 & 0.28 \\
0.21 & 0.22 & 0.19 & 0.38 \\
0.25 & 0.24 & 0.20 & 0.31 \\
0.24 & 0.24 & 0.20 & 0.32
\end{array} \right]
$$

$$
\mathbf{P^{10}} = \left[\begin{array}{cccc}
0.24 & 0.24 & 0.20 & 0.32 \\
0.24 & 0.24 & 0.20 & 0.32 \\
0.24 & 0.24 & 0.20 & 0.32 \\
0.24 & 0.24 & 0.20 & 0.32
\end{array} \right].
$$

Notice that after 10 years, the location of the necklace is independent of its location in year 1. If you are searching for the necklace, the best place to start is Island 4, but the probability it is there is only about 1/3. Alternatively you could camp on an island and wait until the necklace arrives. You are guaranteed that it will return to any island. But, since you must apply for a leave of absence from the world, you'd like to have some idea of how long you must wait. Suppose you would prefer to wait on Island 1. You reconfigure the original transition matrix to be

$$
\begin{bmatrix}
1 & 0 & 0 & 0 \\
4/8 & 0 & 4/8 & 0 \\
0 & 1/8 & 0 & 7/8 \\
3/8 & 3/8 & 2/8 & 0
\end{bmatrix}.
$$

Now you can use the machinery developed for absorbing chains. Specifically you ask for the waiting time to absorption, which is just $\mathbf{Nc}$. Recall

$$\mathbf{N} = (\mathbf{I} - \mathbf{Q})^{-1}$$

where

$$
\mathbf{Q} = \begin{bmatrix}
0 & 0.5 & 0 \\
0.125 & 0 & 0.875 \\
0.375 & 0.250 & 0
\end{bmatrix}
$$

so that

$$
\mathbf{N} = \begin{bmatrix}
1.41 & 0.90 & 0.79 \\
0.82 & 1.80 & 1.58 \\
0.73 & 0.79 & 1.69
\end{bmatrix}
$$

and

$$
\mathbf{Nc} = \begin{bmatrix}
3.10 \\
4.20 \\
3.21
\end{bmatrix}.
$$

If the necklace is currently on Island 2, you should prepare to wait 3.1 years before it gets to Island 1. If it is on Island 3, you must prepare for 4.2 years, and 3.2 if it is on Island 4.

As a second example, consider this modification of the marriage system among the sibs of the Purum. The observed relative frequencies of marriages structured by sib membership, taken from White (1963), is

	Marrim	Makan	Kheyang	Thao	Parpa
Marrim	0	2/23	11/23	10/23	0
Makan	7/23	0	16/23	0	0
Kheyang	1/26	3/26	3/26	10/26	9/26
Thao	0	2/5	1/5	0	2/5
Parpa	7/7	0	0	0	0

The rows report the frequencies of males marrying females belonging to the column sib. For example, there were 23 marriages by Marrim males. Of these, 2 were with Makan females, 11 with Kheyang females, etc. In order to consider this as a Markovian transition matrix, imagine that all offspring of a marriage belong to the mother's sib. This is equivalent to a row selecting a column and then being transformed into a member of the column. This works if it is assumed that each marriage produces at least one son.

Consider a male Marrim who marries a Makan female. Their son then marries a Kheyang female. Their son marries a Thao female. Their son marries a Parpa female. And their son marries a Marrim female. So we might be interested in the number of generations expected for the males of Marrim to be restored to Marrim. The analytic trick introduced in the last example will produce the desired result.

First find

$$\mathbf{N} = (\mathbf{I} - \mathbf{Q})^{-1} = \begin{bmatrix} 1.3 & 1.12 & 0.43 & 0.56 \\ 0.43 & 1.61 & 0.62 & 0.81 \\ 0.61 & 0.77 & 1.30 & 0.79 \\ 0 & 0 & 0 & 1.00 \end{bmatrix}$$

and

$$\mathbf{Nc} = \begin{bmatrix} 3.42 \\ 3.47 \\ 3.46 \\ 1.00 \end{bmatrix}.$$

Since Marrim males do not marry Parpa females the last entry is not meaningful. If a Marrim male initially marries a Makan female, in about 3.42 generations the lineage will again produce a Marrim son.

4.3.3. Information and Markovian Dependence

Applications of information theory continue to amplify and extend studies quite remote from the original purpose (Demetrius, 1974, 1975, 1976). An interesting way of characterizing Markovity was developed by Lila Gatlin (1972) in studying the structure of information in DNA. The technique seems particularly well suited to the study of behavioral programming. At least that is the context to be presented here.

Behavioral programming refers to the relative rigidity of a behavioral sequence. Psychologists use the term scripting to refer to a rigidly sequenced series. One could equally say that the sequence is choreographed. Most primate, including human, behavior is not rigidly programmed at the level of the ethogram. For example, in the industrialized world being left-handed is a matter of indifference except for the nuisance of fitting into a right-handed world. There is no script for responding to left-handedness. "Lipsmacking" is a greeting among Beach troop baboons (Ransom, 1981). So are "tongue protrusion," "jaw-clapping," "ear-flattening," and "eyes narrowed" greetings.

There is no apparent preference in baboon etiquette for any one of these and "usually" lipsmacking, tongue protrusion, and jaw-clapping are combined. Note that even the more generalized event, greeting, may not occur under conditions where it would be appropriate. That is, of the specific behaviors recognized as greetings, it may occur that none are used when two animals come into proximity. In some human rituals, for example a religious cere-mony, the sequence of events is clearly defined and deviation from the script may be sufficient to obviate its benefits. The rigidity of military discipline in a peace-time unit in garrison is frequently cited by Marines (at least) to justify their desire to get back to "the field."

These are examples of a rather simple kind of programming designed by humans for human consumption. Nature is more subtle.

Before considering sequences of behavior it is necessary first to define the state space. This is a troublesome matter. The difficulty is probably sufficient to account for the general poverty of behavioral analysis. You should be clear that the problems are not formal, or even theoretical. I'm not sure what they may be, but the lack of standardization of operations is certainly a contributing factor.

We wish to consider only social behavior so for a group of n individuals, there are n actors and $n - 1$ recipients or targets. The actor is different from the target by definition. Let the behavior set contain k elements. An element of the state space is specified by an actor, a behavior, and a target, so the state space contains $j = nk(n - 1)$ elements. Clearly this space gets large very quickly and the size itself creates observational and other kinds of problems in the execution of research. Notice that each element, $S_i : i = 1, 2, \ldots, j$, is a specific social event. The measure of the total potential randomness in this space is entropy. Maximum entropy occurs when all elements of the space are equally likely. For this space $P(S_i) = p_i = 1/j$ for all i and the maximum entropy is defined to be

$$\begin{aligned} H_0^{(1)} &= \log_2(j) \\ &= -\log_2(p_i). \end{aligned} \tag{4.26}$$

(For the remainder of this section all logarithms will be to base 2. If you cannot obtain these directly, you might refer to the Appendix for the method of con-verting from other bases to base 2.)

When the elements are not equiprobable the entropy is given by

$$H_1^{(1)} = -\sum_i p_i \log_2(p_i) \tag{4.27}$$

which is just the expected value of $\log_2(p_i)$. This expression is due to Shannon. The reduction in entropy produced by deviation from equiprobability is

$$D_1 = H_0^{(1)} - H_1^{(1)}. \tag{4.28}$$

If the elements of S are occurring independently then

$$P(S_m | S_i) = P(S_m)$$

for all m. An extreme case of divergence from independence would be the sequence

$$S_1 S_1 S_1 \cdots S_2 S_2 S_2 \cdots S_i S_i S_i \cdots S_j S_j S_j.$$

Here the $P(S_i|S_i)$ are all high, the $P(S_{i+l}|S_i)$ are low, and the $P(S_{i+l}|S_i) = 0$ for $l \geq 2$. Another extreme case of divergence from independence is the sequence

$$S_1 S_2 S_3 \cdots S_j S_1 S_2 \cdots S_j S_1 S_2 \cdots S_j.$$

Here the $P(S_{i+1}|S_i) = 1$ and all others are 0. You should note that in both these cases $P(S_1) = P(S_2) = \cdots = P(S_j) = 1/j$ yet divergence from independence is extreme.

We need an expression which will quantify the amount of departure from independence for less than extreme cases. Any observed sequence of social behaviors is a finite linear ordering of sampled elements from S. Define a new state space containing pairs of behavioral events in a sequence. For example, if the sequence $(S_2 S_1 S_3)$ is observed, the elements of the new state space are $(S_2 S_1)$ and $(S_1 S_3)$. The entropy of this state space is

$$H_1^{(2)} = -\sum_{i,j} P(S_i S_j) \log_2(P(S_i S_j)) \tag{4.29}$$

where $j = i + 1$. $H^{(2)}$ is a maximum when the elements are pairwise independent, that is when

$$P(S_i S_j) = P(S_i) P(S_j).$$

(In the remainder of this discussion, the base of all logarithms is assumed to be 2 so the subscript will be dropped.) Entropy under this condition is

$$H_0^{(2)} = -\sum p_i p_j \log(p_i p_j)$$
$$= -\sum \sum p_i p_j \log(p_i) - \sum \sum p_i p_j \log(p_j).$$

Now noting that $\sum p_i = 1$ this becomes

$$H_0^{(2)} = -\sum p_i \log(p_i) - \sum p_j \log(p_j) \tag{4.30}$$

and since S_i and S_j are formally equivalent

$$H_0^{(2)} = 2H_1^{(1)}. \tag{4.31}$$

When the elements are not pair-wise independent then

$$P(S_i S_j) = P(S_i) P(S_j|S_i)$$

which results in entropy

$$H_1^{(2)} = -\sum p_i p_{ij} \log(p_i p_{ij}) \tag{4.32}$$

where $p_{ij} = P(S_j|S_i)$. Then because $\sum_j p_{ij} = 1$

$$H_1^{(2)} = -\sum [p_i \log(p_i) + p_{ij} \log(p_{ij})]$$
$$= H_1^{(1)} - \sum p_{ij} \log(p_{ij}). \tag{4.33}$$

The second term on the right of (4.33.) will be called $H_M^{(2)}$, or H-Markov. (Note that only one step transitions are involved. Clearly this could be extended.) Then

$$H_1^{(2)} = H_1^{(1)} + H_M^{(2)}. \tag{4.34}$$

The divergence from independence is

$$
\begin{aligned}
D_2 &= H_0^{(2)} - H_1^{(2)} \\
&= H_1^{(1)} - H_M^{(2)}.
\end{aligned} \tag{4.35}
$$

Shannon defined redundancy as

$$R = 1 - H_M^{(2)}/\log(j) \tag{4.36}$$

and since

$$\log(j) - H_M^{(2)} = D_1 + D_2$$

then

$$R \log(j) = D_1 + D_2. \tag{4.37}$$

The common understanding of redundancy is repetition. That which is defined above is a much more subtle and effective means of avoiding errors in communication. Redundancy measures the amount by which entropy has been reduced from its maximum; it measures the effect of all the ordering, constraints, rules, etc. that constitute the system (Gatlin, 1972). In order to investigate the structure of redundancy, define two indexes of relative contribution to the reduction of entropy (*ibid*):

$$RD1 = D_1/(D_1 + D_2) = D_1/(R \log(j)) \tag{4.38a}$$

$$RD2 = D_2/(D_1 + D_2) = D_2/(R \log(j)) \tag{4.38b}$$

Two qualitatively different strategies for increasing redundancy are, thus, available: (1) increase D_1 relative to D_2, i.e. change the relative frequency of the simple events; or (2) increase D_2 relative to D_1, i.e. make some simple events conditional on some others. The vertebrates achieved a high level of redundancy in their DNA by holding D_1 constant and increasing D_2 (*ibid*).

Increasing redundancy in information, whether behavioral or chemical, by increasing $RD2$ is the means of optimizing the conflicting elements of variety versus reliability (*ibid*). Ritualization, for example, increases $RD2$.

In order to illustrate these concepts, we observed the social behavior of a group of four deBraza monkeys (*Cercopithecus neglectus*) at the Stanley Park Zoo in Vancouver. The exercise was conducted for the purpose of determining the degree of structure, choreography, in "touching." As defined, touching is a directed action, that is an act is "B touches D." This excludes incidental contact. The decision was made to exclude the adult male. The early observations indicated that he was not part of the tactile dimension of social structure. This left three animals—two adult females, here identified as B and C, and a young juvenile, the offspring of B.

Using three animals and a single directed behavior produces a behavior set with 6 elements: (BC, BD, CB, CD, DB, DC). The notation BD means an observation of the event B touches D. The maximum entropy of this set is

$$H_0^{(1)} = \log_2(6) = 2.585.$$

It is clearly the case that the behavioral events in this space are not equally probable. Specifically, over the observation period the following relative frequencies were obtained:

Event	Relative frequency
BC	0.087
BD	0.100
CB	0.061
CD	0.074
DB	0.504
DC	0.170

It will be noticed that about half of the total social activity was the juvenile touching its mother, the event DB. It touched the other female only about 17% of the time. The least frequent event, CB, is the female C touching female B. Only slightly more frequently did C touch D, the juvenile. The entropy of the realized behavior state space, from equation 4.25 is

$$H_1^{(1)} = -\sum p_i \log(p_i)$$
$$= -[(0.087) * (-3.523)$$
$$+ (0.100) * (-3.322)$$
$$+ (0.061) * (-4.035)$$
$$+ (0.074) * (-3.756)$$
$$+ (0.504) * (-0.989)$$
$$+ (0.170) * (-2.556)]$$
$$= 2.096$$

which means that the reduction of entropy in this space due to deviations from equal probability is

$$D_1 = 2.585 - 2.096 = 0.489.$$

The state space of behavioral interaction is specified by the sequence of social acts. An interaction is described by BD, CB which means that B touches D and then C touches B. This produces a state space with 36 elements,

[(BC, BC), (BC, BD), (BC, CB), ..., (DC, DC)] with maximum entropy

$$H_0^{(2)} = \log(36) = 5.170 = 2H_1^{(1)}.$$

The realized social interaction matrix is

	BC	BD	CB	CD	DB	DC
BC	0.03	0.00	*0.00*	0.00	0.01	0.03
BD	0.00	0.01	0.00	0.02	*0.06*	0.00
CB	*0.01*	0.00	0.02	0.00	0.02	0.00
CD	0.00	0.01	0.01	0.02	0.01	*0.02*
DB	0.02	*0.06*	0.01	0.01	0.36	0.04
DC	0.01	0.00	0.02	*0.03*	0.04	0.07

where the reciprocal sequences are italicized. You will note that simple, one step, reciprocity does not seem to characterize the interaction of these monkeys. The entropy of the realized interaction matrix is, from equation 4.29

$$H_1^{(2)} = 3.802.$$

The reduction in maximum entropy which is attributable to one step, Markovian, dependence in the behavior sequence is

$$D_2 = 5.170 - 3.802 = 1.368.$$

The total reduction in entropy from both sources—deviation from equiprobable acts and Markovian dependence in the action sequence—is

$$D_1 + D_2 = 1.859$$

of which the percentage due to Markovity is about 74%.

4.3.4. Games with Strategic Uncertainty

To this point most attention has been given to processes with stochastic uncertainty. Here we shift focus a bit and attend processes which depend on interaction such that each player must select a behavior in consideration of others.

The theory of games has been extensively and successfully applied to economic and military behavior in the modern industrial context. Most attempts to apply the methods to topics of anthropological interest have not been particularly successful. Anthropologists have broadly rejected the theory because of this. But this is, again, blaming the tool for the inadequacy of the carpenter. The reason for the failures seems obvious—the things which motivate behavior in the short term are not universal, they are subject to "cultural" manipulation. People appear to behave irrationally because they do not attempt to maximize the things that economists or generals think they should. In the long run, in the evolutionary context, the concept of human rationality is replaced by that of evolutionary stability (Maynard-Smith, 1982). It then seems irrational to suggest that any organism does not attempt to maximize its fitness in some sense. Otherwise extinction is inevitable.

You may have heard that game theory models "rational behavior." When the goal of the game is to maximize money it is the case that, for a given strategy set, a game theoretic analysis will usually assist in selecting optimal behavior. This goal is so pervasive in the culture of the western world that it seemed reasonable to extend the conceptual structure to other cultural environments—in effect asserting that the pursuit of money is a universal motivator. It is not. It is necessarily the case that all organisms attempt to increase, or at least maintain, their biological fitness. The interesting theoretical problem for anthropology is to develop the logic establishing the relationship of culture to fitness. This is one of three basic problems which evolutionary theory has not resolved.

A primitive concept of game theory is the strategy. This should be considered a kind of behavior. The game itself is expressed as a payoff matrix which shows that amount won or lost by each player under all possible choices of strategy by each. We shall restrict attention to two player games, but this limitation is for convenience only. Another restriction adopted for similar reasons is that whatever one player wins, the other loses. In the parlance of game theory, these are two-person zero-sum games.

The zero-sum condition means that the games of interest here are competitive. Hurwicz (1968) illustrates the concept by considering Columbus' problem. The structure of the decision confronting him was as

| | | Distance to land | |
		Land near	No land near
Columbus' decision	Turn back	Probable later disappointment	Life saved
	Keep going	Prospect of glory	Prospect of death

Here Columbus is playing against nature. Suppose that he assigned payoffs in terms of "satisfaction units"

	Land near	No land near
Turn back	−50	20
Keep going	100	−1000

Also he has decided that the probability that land is near is 3/4. His expected satisfaction is

Turn back $3/4(-50) + 1/4(20) = -32.5$
Keep going $3/4(100) + 1/4(-1000) = -175.0.$

Clearly the potential lost satisfaction if he keeps going is much greater than if he turns back. It would have been rational, as well as prudent, to turn back. In fact, if this payoff matrix is close to reality then he would have required that the probability of land near be 9/10. But, he muses, perhaps the fear of death is too great and the value of the prize too small. Suppose the payoff matrix is

	Land near	No land near
Turn back	-1000	20
Keep going	500	-500

Now his expected satisfaction is

Turn back $3/4(-1000) + 1/4(20) = -745$
Keep going $3/4(500) + 1/4(-500) = 250$

and his rational choice is to continue.

Daly and Wilson (1983) present data from a study of 19th century Mormon households which will serve as an example of a cooperative game. In Table 4.2 the (approximate) results of several family structures are presented.

Table 4.2. Polygyny and Fertility among Nineteenth-Century Utah Mormons. (Adapted from Daly and Wilson, Fig. 11-2)

		Female payoff (offspring) Wife		
		First	Second	Third
Male payoff (off-spring)	Monogamous households	7	4	—
		7	8	—
	Polygynous 2-wives	9	6	—
		9	17	—
	Polygynous 3-wives	7	6.5	6
		7	14	20

N.B. Offspring of wife above, of husband below.

Clearly if the objective is maximizing offspring, the male strategy "get as many wives as possible" is best. The best female strategy is to be the "first wife in a 2-wife household." This is not a zero-sum game because the payoff to one partner is also a payoff to the other. This is a cooperative, not a competitive game.

Luce and Raiffa (1952) describe the following game:

During the World War II battle for New Guinea, the American commander, General Kenny, had intelligence that a Japanese supply convoy would arrive. The convoy had the choice of a northern route with poor visibility, or a southern route which would be clear. Kenny had the choice of concentrating his reconnaissance aircraft on one or the other of these routes. Kenny's staff estimated the number of days of bombing time for the different choices to be

		Japanese strategy	
		North	South
Kenny's search	North	2	2
strategy	South	1	3

Note that if Kenny chose the northern route, he could expect a minimum of 2 bombing days for each possible choice by the Japanese commander. If Kenny chose the southern route, he could expect a minimum of 1 bombing day. So Kenny chose to concentrate his reconnaissance on the northern route. The Japanese commander would have reasoned similarly noting, that if he went north the maximum bombing days would be 2 whereas if he went south the most was 3. So the Japanese commander also chose the northern route. This became the Battle of the Bismark Sea and was a disastrous defeat for the Japanese. For the Japanese commander, north was the least bad of two costly choices.

In this game there is an entry which is simultaneously the minimum of its row and the maximum of its column. The north-north entry is such an entry. This is called a saddle point for the game. When a saddle point exists, then the best choice for the row player is the row containing the point, and likewise the best choice for the column player is the column containing the point. The payoff at this point is the value of the game. (Notice that by convention, payoff matrices will contain payoffs to the row player. This is acceptable for a zero-sum competitive game structure.) The game is strictly determined and the strategies for both players do not change—the strategy for both is called a pure strategy.

Now consider the following game. There are two strategies available to individuals in a population, Hawk and Dove. Hawks are always aggressive and Doves always passive. Evolution in the population occurs as a result of encounters between individuals according to the payoff matrix

	Hawk	Dove	Minimum
Hawk	−1	2	−1
Dove	−2	1	−2
Maximum	−1	2	

Consider the row player first. If he plays Hawk the least he will receive (yield) is −1, whereas playing Dove will return a minimum −2. Clearly he will opt for −1 rather than −2. The column player receives a maximum of 2 playing Hawk and playing Dove he yields 2. (Note that the entries in the matrix are expressed in terms of the row player's winnings, so an entry of −2 is a win for the column player and 2 is a loss.) He will play Hawk because this produces a win of 2 units for him whereas Dove results in his giving 2 to the row player. Since the row minimum and the column maximum coincide at Hawk-Hawk, this is the saddle point.

Games with pure strategies do not enter the research literature. Once you can locate a saddle point—the maximum of the row minimums and the minimum of the column maximums—you have nearly everything of interest for strictly determined games. The games that have great theoretical interest in evolutionary studies are those requiring mixed strategies by the players. Consider the game with payoff matrix

$$\begin{array}{ccc} & S_1 & S_2 & \text{Minimum} \\ S_1 & \begin{bmatrix} -1 & 5 \\ 2 & -3 \end{bmatrix} & & \begin{array}{c} -1 \\ -3 \end{array} \\ \text{Maximum} & 2 & 5 \end{array}$$

Both players should prefer S_1 but note that the entry $S_1 S_1$ is not a saddle point. It is not simultaneously the minimum of its row and the maximum of its column. Suppose the game is being played repeatedly and that the row player consistently selects S_1 hoping to win 5. The column player counters by selecting S_1 and receiving 1. If the row player should play S_2 consistently then the column player will choose S_2 and win 3. You might expect that some mix of S_1 and S_2 would be the best way for both players to play the game. Part of the time play S_1, and play S_2 for the remainder. This possibility raises some additional problems, however. Specifically, how often should you play S_1? And on any given play how do you decide between S_1 and S_2?

Suppose, to fix the concept, both the row and column players decide separately to play each available strategy equally frequently. In the case at hand, this means play S_1 on 1/2 of the plays and S_2 on the remainder. Now the problem is how do you choose a strategy for the next game? If the duration of the game is known in advance you might decide to play S_1 on the first 1/2 and S_2 on the second 1/2. But your opponent would figure that out rather quickly. What you need is a device which will select S_1 randomly 50% of the time overall. Tossing a coin is an adequate approximation. If this is done, the row player expects to win

$$(-1)(1/2) + (2)(1/2) = 1/2$$

when the column player picks S_1 and

$$1/2(5) + 1/2(-3) = 1$$

when the column player picks S_2. The column player, using the same mix, expects

$$1/2(-1) + 1/2(5) = 2,$$

which is a loss to him, when the row player picks S_1, and

$$1/2(2) + 1/2(-3) = -1/2$$

when the row player picks S_2. In the long run the row player expects to win

$$1/2(1/2) + 1/2(1) = 3/4$$

and the column player expects to lose

$$1/2(2) + 1/2(-1/2) = 3/4.$$

Now let us generalize these results a bit. Let the payoff matrix be

$$\begin{bmatrix} a_{11} & a_{12} \\ a_{21} & a_{22} \end{bmatrix}$$

and the row player's mix will be the row vector

$$[r_1 \quad r_2]$$

where $r_1 + r_2 = 1.0$. The expected return to the row player is

$$[r_1 \quad r_2]\begin{bmatrix} a_{11} & a_{12} \\ a_{21} & a_{22} \end{bmatrix} = [r_1 a_{11} + r_2 a_{21} \quad r_1 a_{12} + r_2 a_{22}]. \qquad (4.39)$$

Now let the column player's mix be the column vector

$$\begin{bmatrix} c_1 \\ c_2 \end{bmatrix}$$

where $c_1 + c_2 = 1.0$. He expects to win

$$\begin{bmatrix} a_{11} & a_{12} \\ a_{21} & a_{22} \end{bmatrix} \cdot \begin{bmatrix} c_1 \\ c_2 \end{bmatrix} = \begin{bmatrix} c_1 a_{11} + c_2 a_{12} \\ c_1 a_{21} + c_2 a_{22} \end{bmatrix}. \qquad (4.40)$$

The value of the game is

$$V = [r_1 \quad r_2]\begin{bmatrix} a_{11} & a_{12} \\ a_{21} & a_{22} \end{bmatrix} \cdot \begin{bmatrix} c_1 \\ c_2 \end{bmatrix} \qquad (4.41)$$

$$= [r_1 c_1 a_{11} + r_1 c_2 a_{12} + r_2 c_1 a_{21} + r_2 c_2 a_{22}].$$

You should convince yourself that this result holds for arbitrary dimensions, i.e. any number of strategies, of the payoff matrix. For example, consider the game with payoff matrix

$$\begin{bmatrix} 2 & 0 & -1 \\ -1 & 3 & 4 \end{bmatrix}.$$

The row player will play $[1/2 \quad 1/2]$ and the column player wishes to know which of the following mixes he should play. (Note that the column player is interested in making the value of the game as small as possible.)

$$\mathbf{A} = \begin{bmatrix} 0.6 \\ 0.3 \\ 0.1 \end{bmatrix} \quad \text{or} \quad \mathbf{B} = \begin{bmatrix} 0.3 \\ 0.3 \\ 0.4 \end{bmatrix}.$$

Using mix $\mathbf{A}$ the value of the game is

$$\mathbf{V_A} = [0.5 \quad 0.5] \begin{bmatrix} 2 & 0 & -1 \\ -1 & 3 & 3 \end{bmatrix} \begin{bmatrix} 0.6 \\ 0.3 \\ 0.1 \end{bmatrix} = 0.9$$

and using mix $\mathbf{B}$ it is

$$\mathbf{V_B} = [0.5 \quad 0.5] \begin{bmatrix} 2 & 0 & -1 \\ -1 & 3 & 4 \end{bmatrix} \begin{bmatrix} 0.3 \\ 0.3 \\ 0.4 \end{bmatrix} = 1.2.$$

So it is to his advantage to play mix $\mathbf{A}$.

The row player suspects he could do better and becomes interested in the question of finding the best possible strategy mix for the game. He wishes to find that mix of strategies, against an opponent playing his best mix, which makes the value of the game a maximum.

In considering the row player's problem of finding a best strategy mix it is useful to distinguish 2×2 games—both players have two available strategies—from larger ones. The main reasons for this are that the 2×2 game is simpler to analyze and is important in many theoretical applications. The payoff matrix for this special case will be

$$\begin{bmatrix} a & b \\ c & d \end{bmatrix},$$

the strategy mix for the row player is $[x \quad 1 - x]$, and for the column player

$$\begin{bmatrix} y \\ 1 - y \end{bmatrix}.$$

Now assume the column player chooses S_1. The expected payoff to the row player is

$$ax + c(1 - x) \geq v. \tag{4.42}$$

Since the game has a guaranteed solution, its value is v when the column player plays S_1. The row player expects to receive at least the v. Hence the inequality. And when the column player picks S_2, the row player expects

$$bx + d(1 - x) \geq v. \tag{4.43}$$

The column player has the expectation of at most v for the game. So when the row player selects S_1 this is

$$ay + b(1 - y) \leq v \tag{4.44}$$

and when he plays S_2 the column player expects

$$cy + d(1 - y) \leq v. \tag{4.45}$$

Since the game has no saddle point

$$0 < x < 1 \quad \text{and} \quad 0 < y < 1.$$

The game is solved when we have values for x, y, and v.

First consider the row player's inequalities. As a convenience, these are expressed as equalities in order to make the solution simpler. Then since they both equal the same thing

$$ax + c(1 - x) = bx + d(1 - x)$$

which may be solved for x

$$x = d - c/((a + d) - (b + c)). \tag{4.46}$$

It is equally easy to solve the equalities for the column player. Write

$$ay + b(1 - y) = cy + d(1 - y)$$

and solve for y

$$y = d - b/((a + d) - (b + c)). \tag{4.47}$$

There are several things you should note about equations 4.46 and 4.47. Since both x and y are probabilities, they are both positive. In order for 4.46 to be positive it must be the case that either

$$[(d > c) \text{ and } (a + d) > (b + c)]$$

or

$$[(d < c) \text{ and } (a + d) < (b + c)]$$

is true. Similarly for 4.47 to be positive, either

$$[(d > b) \text{ and } (a + d) > (b + c)]$$

or

$$[(d < b) \text{ and } (a + d) < (b + c)]$$

is true. These conditions specify the relative magnitudes of the payoffs which characterize a game with no saddle point:

1. $a > d > b > c$.
2. $d > a > b > c$
3. $a > d > c > b$
   etc.

that is, when both payoffs in one of the diagonals of the matrix is greater than either of the payoffs in the other diagonal.

The value of the game is

$$v = ad - bc/((a + d) - (b + c)). \tag{4.48}$$

This basic analytic procedure also works for larger games. The number of conditions on the solution make the process unwieldy with even small strategy sets so we shall not go any further with this.

For example consider the game with payoff matrix

$$\begin{bmatrix} 3 & -1 \\ -2 & 1 \end{bmatrix}$$

for which we wish to find optimal strategies and the value of the game. First find

$$D = a + d - c - b$$

$$3 + 1 - (-2) - (-1) = 7.$$

The optimal mix for the row player is

$$x = (1 - (-2))/7 = 3/7$$

$$1 - x = 4/7$$

from equation 4.46 and for the column player it is

$$y = (1 - (-1))/7 = 2/7$$

$$1 - y = 5/7.$$

If both players play their optimal strategy mix, the payoff is

$$v = ((3)(1) - (-2)(-2)(-1))/7 = 1/7,$$

that is the row player will gain 1/7 and the column player will lose that amount.

Now let us treat the problem of the optimal strategy mix from another perspective. The technique uses a graphical approach and so is limited to two players using an arbitrary number of strategies. It is convenient to have all payoffs be positive. Adding 3 to all entries we obtain

$$\begin{bmatrix} 6 & 2 \\ 1 & 4 \end{bmatrix}.$$

Find

$$D = (a + d) - (b + c)$$
$$= (6 + 4) - (1 - 2) = 7$$

and

$$x = (4 - 1)/7 = 3/7$$
$$y = (4 - 2)/7 = 2/7$$

which is the same mix already obtained. The difference is in the value. Calculate

$$((6)(4) - (1)(2))/7 = 22/7$$

but recall that $3 = 21/7$ was added to all entries so

$$v - 3 = 22/7 - 21/7 = 1/7$$

and all is as it should be. So nothing is lost by having all the entries be positive. The inequalities 4.42 and 4.43, after dividing through by v, become

$$a(x/v) + c((1 - x)/v) \geq 1$$
$$b(x/v) + d((1 - x)/v) \geq 1.$$

Now let

$$z_1 = x/v$$
$$z_2 = (1 - x)/v$$

and the inequalities become

$$(1) \ 6z_1 + z_2 \geq 1$$
$$(2) \ 2z_1 + 4z_2 \geq 1.$$

Each of these inequalities defines a part of the half plane, their conjunction defines a region of solutions, and, by a very clever theorem, their intersection will be the value of the game. Let us solve these inequalities for z_1 and z_2. First, to eliminate z_2 multiply (1) by 4 and subtract the result from (2):

$$
\begin{array}{llll}
(2) & 2z_1 & + 4z_2 \geq & 1 \\
\text{minus} \quad (3) & 24z_1 & 4z_2 \geq & -4 \\
\hline
& -22z_1 & \geq & -3
\end{array}
$$

which can be solved for z_1

$$z_1 \geq 3/22,$$

then substituting into (2) to find z_2

$$2(3/22) + 4z_2 \geq 1$$
$$z_2 \geq (16/22)/4 = 4/22.$$

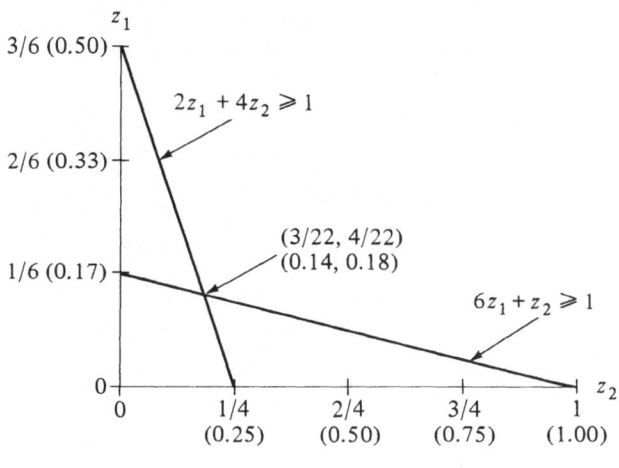

Figure 4.5

Now with these values of the vector **z**, using the fact that

$$v = 1/(z_1 + z_2)$$

we find

$$v = 22/7$$

which is the value obtained earlier. To find the optimal strategy mix for the row player we make use of

$$x_1 = vz_1 = (22/7)(3/22) = 3/7$$
$$1 - x_2 = vz_2 = (22/7)(4/22) = 4/7$$

as they should.

The solution of simultaneous equations can be quite tedious. It is useful to have an alternate trick. In Figure 4.5 the inequalities are plotted. The values of z which optimize the strategy mix at the intersection of the two lines. The lines are obtained as follows. First, in inequality (1), set $z_2 = 0$ and find $z_1 \geq 1/6$; then set $z_1 = 0$ and find $z_2 \geq 1$. These are the coordinates of the boundary of the first inequality as it passes through the axes. Second, in inequality (2) set $z_2 = 0$ and find $z_1 \geq 1/2$; then set $z_1 = 0$ and find $z_2 \geq 1/4$. From the graph one may read that $z_1 = 0.14 \sim 3/22$ and $z_2 = 0.18 \sim 4/22$.

A particular game called The Prisoner's Dilemma has been broadly applied in a variety of contexts (Luce and Raiffa, 1957; Jones, 1980; Eigen and Winkler, 1981; Maynard-Smith, 1982). The origin of the name is attributed to A.W. Tucker who described the game as follows:

> Two suspects are taken into custody and separated. The prosecutor is certain that they are guilty of a specific crime, but he does not have adequate evidence to convict them at a trial. He makes the following proposal to each prisoner. If

they both confess each will serve 8 years, if neither confesses each will serve 1 year. If one confesses and the other does not the confessor will serve 3 months and the other 10 years. (Luce and Raiffa, 1957)

Here a prisoner is said to "defect" if he confesses, and to "cooperate" if he refuses. The general payoff matrix is usually presented as

	Cooperate	Defect
Cooperate	R (reward)	S (sucker)
Defect	T(temptation)	P (punishment)

with $T \geq R \geq P \geq S$ and $R \geq (S + T)/2$. When $T \geq R$ it pays to Defect if the other Cooperates. If $P \geq S$ it pays to Defect if the other Defects. So Defect is the optimal strategy for both players (Axelrod and Hamilton, 1981). Hamilton's definition of inclusive fitness (1964), and Maynard-Smith's concept of Evolutionary Stable Strategy, ESS, (1973) allowed the extension of game theory into evolutionary theory.

The game which Maynard-Smith (1982) analyzes, with profound implications for behavioral and phenotypic evolution, is the following: There are two strategies

Hawk: escalate aggression and continue to fight until injured or the opponent retreats.

Dove: display and retreat immediately if the opponent escalates.

The value, in terms of increased fitness, of the resource is v. The cost, in terms of lost fitness, is c. The payoff matrix is

$$
\begin{array}{c}
 \\
\text{Hawk} \\
\text{Dove}
\end{array}
\begin{array}{cc}
\text{Hawk} & \text{Dove} \\
\left[\begin{array}{cc}
(v - c)/2 & v \\
0 & v/2
\end{array} \right] .
\end{array}
$$

Clearly Dove is not an optimal strategy. Hawk is optimal if $v \geq c$, if the value to be won is greater than the cost. If $v \leq c$ neither Hawk nor Dove is optimal. The strategy mix with $P(\text{Hawk}) = v/c$ is optimal.

As mentioned earlier, the work of Hamilton (1964) and Maynard-Smith and Price (1973) allowed for the extension of game theory in the theoretical development of behavioral evolution. The Prisoner's Dilemma structure has been conceptually central to much of this recent explosion. This is an excellent conceptual model of the behavior of organisms under strict Darwinian evolution—selfishness is predicted and observed. But so is cooperation observed and this was a serious difficulty for evolutionary theory until Hamilton showed that inclusive fitness, not individual fitness, is maximized in evolution. But the problems were not over. Maynard-Smith and Price showed that some behavioral strategy mixes are strongly selected and not subject to modification; the evolutionary stable strategy, ESS, and theoretical work on the iterated Prisoner's Dilemma by Axelrod and Hamilton (1981) was required to show how cooperation could evolve as an ESS.

The structure has been used very effectively to describe parent-offspring conflict (Trivers, 1974), sex ratio (Maynard-Smith, 1982), coalitions among male baboons (Packer, 1977), and many other specific behaviors.

CHAPTER 5

Topics in Hypothesis Testing

5.1. Introduction

Once more recall the conditional statement and the scientific argument which is being evaluated:

$$if \ [not \ H] \ then \ [probably \ not \ P]$$
$$[not \ [not \ P]]$$
$$thus \ [H].$$

This is simply a reminder. The hypothesis has produced a prediction and we are confident that if the hypothesis is false then the prediction will not be observed. In this chapter we shall be focussing attention on the prediction. Specifically we shall ask whether observations in the world are consistent with the prediction. The central part of any scientific program, however, is the theory and the hypothesis that produces the prediction.

From the structure of the scientific program which has been presented the specific hypothesis which is to be tested is [*not* P]. If the observations are consistent with P, that is, the prediction occurs, then we shall reject [*not* P]. The result is the argument structure "deny the consequent," which, then, deductively results in the negation of the antecedent, that is [*not* H]. Recall that the observations will actually support the research hypothesis only when the consequent in the conditional statement follows deductively from the antecedent. In the social sciences this relationship does not commonly obtain.

In the conditional statement for the scientific program the consequent is [*not* P]. This is the origin of a concept which is frequently puzzling to students new to the study of hypothesis testing. It is called the null hypothesis in the literature. Recalling the logical structure of the program, the concept should be clear. If the research hypothesis has produced a prediction by formal argument, then the null hypothesis is conceptually routine. Otherwise considerable attention must be given to the formulation of this hypothesis since the overall validity of the project depends very heavily upon it. Unfortunately the statistical tests to be presented are indifferent to the quality of the hypothesis. The techniques will test the fit of data to guesses, hunches, speculations, musings, and legitimate hypotheses. This indifference opens the flood gates to all manner of random shots. A class of ready examples is called "spurious" tests/results/correlations. For example, it is the case that where the stork population is high, so is the birth rate. A "hypothesis" about the involvement of storks in human births could be, and has been, humorously (I trust) supported. It is possible to be very smug and complacent because we know a great deal about conception and birth and storks are not included. (You should note that condition 2 could not be satisfied—*if* [*not* storks] *then* [probably *not* high birth rate].) It is essential that you understand that when the scientific program is short circuited, or abandoned entirely as some would have it today, then you run the risk of deciding that storks bring babies, or the equivalent. Here is a fundamental principle that you should carve on your wall:

> The value (worth) of a statistical test of an hypothesis resides entirely in the quality of the hypothesis.

The less formal the argument generating the prediction, the more attention must be given to the null hypothesis. Note that a statistical test will usually produce a decision about the null hypothesis—specifically one may (1) reject the null hypothesis and accept the alternative, or (2) fail to reject the null hypothesis. An option which is always available, but rarely accepted, is to suspend judgment. When the data are more variable than was assumed in the design of the experiment, the result will be equivocal and decision must be foregone.

The argument structure at the beginning may be used to illustrate the concept. Since we wish to accept the theoretical hypothesis, we must observe [*not* [*not* P]] = P. So [*not* P], that is, the negation of the prediction, must fill the role of the null hypothesis. Note that this is a goal oriented step. The goal is to reject the negation of the prediction. This is assuredly not a divine objectivity. All resources will be marshalled in the service of the goal. But if you cheat—and many have either innocently or malevolently—you risk your reputation. This commodity is the only item of professional value that many of us have, and to risk it is unthinkable. So we don't cheat. Fortunately the rules are clear and well defined, so procedural catholicity is the warrant against fraud.

How to Be Wrong

Suppose that you predict Armageddon. If it occurs, then the theory may seem more credible—as it does to Polya (1954)—but you may not logically claim support for the hypothesis on the basis of good inductive argument. If it does not occur, then what? How do you decide that the event has not occurred/is not occurring/will not occur? The short answer is that it can't be done. (This is also the long answer.) The prediction must say something like "Armageddon next Thursday before noon" or you may never be able to decide. This is generally true. The prediction of an event, without specifying its location in time or space or both, is not enough to allow a decision. Even more generally, the prediction of any change of state must be located in space and/or time.

The theoretical hypothesis is a model of the frequency of an event (Hoel, 1962, 1946). It may be either true or false, and the data may or may not reject [not P]. These possibilities mean there are two ways to be wrong: (1) the hypothesis is false and [not [not P]] is observed, or (2) the hypothesis is true and [not P] is observed.

Outcome of experiment or observation—the decision	Theoretical hypothesis	
	H True	H False
not [not P]	Correct decision	Type I error
[not P]	Type II error	Correct decision

A type I error is committed when a false hypothesis is accepted or, more properly, not rejected; and a Type II error is committed when a true hypothesis is rejected. Note well that the outcome of the experiment may support either the prediction or its negation regardless of whether the hypothesis is true or false. The cause of each error is an interaction between a theoretical description of an event (HYPOTHESIS), and data (PREDICTION). The theory, recall, describes the way the world is thought to be. The prediction asserts that some observations will be obtained if the world is that way. If the experiment is well designed, the data should inform about whether the predicted observations were obtained. In order that they actually do so, careful attention must be given to their acquisition. This is because the observations are a finite sample from a potentially infinite population of observations which could have been obtained. The theory should also produce some expectation about variability in order that the number of observations required may be fixed in advance.

A Decision Problem

For example, and to fix the concepts, assume that a fair coin is to tossed. A fair coin, by definition, produces heads with probability 1/2. You do not know whether the coin is fair or biased. This knowledge would be very useful, for you intend to bet on it. Clearly if you bet as though the coin is biased when in fact it is fair, or the reverse, you will lose your shirt. For now, assume that observations are free. If you toss the coin a few times for the purpose of deciding whether it is fair or not, there is a calculable non-zero probability that you will decide it is biased no matter what criterion you use. Suppose that you toss it 10 times and if you get more than 7 or fewer than 3 heads you will decide the thing is biased. The probability of this outcome—either 0, 1, 2, 3, 7, 8, 9, 10 heads—is about 0.34 if the coin is fair. In this case you have the prior expectation, from long experience with a variety of coins, that it is fair. So H is [the coin is fair]. If the probability of a head is greater than or equal to 0.8, or less than or equal to 0.2, then for your purposes it is biased.

From these considerations we get, e.g.

$$\textit{if } [\textit{not } P(\text{head}) \geq 0.8] \textit{ then } [\text{probably}$$
$$\textit{not } (7 \textit{ or } 8 \textit{ or } 9 \textit{ or } 10 \text{ heads})].$$

(Recall that P is the antecedent in a conditional statement and P($\cdot$) is the probability operator.) Fewer than 7 heads would result in your rejection of that part of the null hypothesis which specifies a probability greater than 1/2. A strictly symmetrical argument would allow the rejection of the part specifying a probability less than 1/2. (You may be troubled by the fact that the probability need not be as great as 0.8 or as small as 0.2 for the coin to be biased. For now accept that this is the difference that makes a difference to you.) Now we have the conjunction of two conditions

$$(\textit{if } [\textit{not } (P(\text{head}) \geq 0.8)] \textit{ then } \text{probably } [\textit{not } (\geq 7 \text{ heads})])$$

and

$$(\textit{if } [\textit{not } (P(\text{head}) \leq 0.2)] \textit{ then } \text{probably } [\textit{not } (\leq 3 \text{ heads})]).$$

The negation of the prediction in the first conditional is

$$\textit{not } [\textit{not } (\geq 7 \text{ heads})] = (< 7 \text{ heads})$$

and for the second it is

$$\textit{not } [\textit{not } (\leq 3 \text{ heads})] = (> 3 \text{ heads}).$$

An outcome resulting in fewer than 7 heads allows acceptance of the hypothesis

$$\textit{not } (P(\text{head}) \geq 0.8) = P(\text{head}) < 0.8.$$

And an outcome with more than 3 heads allows acceptance of $P(\text{head}) \geq 0.2$. The conjunction of these is $0.2 < P(\text{head}) < 0.8$ which for your purposes constitutes a fair coin.

As we saw earlier the probability of accepting the alternative—[coin is biased]—when in fact the coin is fair is about 0.34. This is a Type II error. Suppose that instead of 10 sample observations you decide to take 20. Now the probability of a Type II error is about 0.002 if a proportional rejection region is used—i.e. less than 4 or more than 16 heads. Clearly this is much more comfortable and you may be tempted to conclude that increasing the number of observations is the answer to all research problems. If so, you are not alone, but you are certainly wrong.

Suppose you have $21 and each test observation costs $1. If you take 20 then you've only $1 left to play a game with a probability of 1/2, you have decided, of winning on any given trial. You will likely be wiped out rather early. There is a more serious objection, however, to increasing the sample size mindlessly. It is possible for example, to accumulate enough observations to discriminate between a probability of 1/2 and one of 0.495. And it is conceivable that this difference is important, but for the level of most theory in anthropology, it is only conceivable. One should not make any more observations than are required to reach a decision about a hypothesis for a given level of confidence. This is always a fraction of the number required to reach that level of confidence in deciding that the probability is within an interval of arbitrary width.

The problems of choosing a rejection region and the number of observations to be made are strongly interrelated. The clear solution is avilable only when the cost of each kind of error is known. For example testing a drug for the treatment of cancer clearly involves different costs than testing a drug for the treatment of hangnails. The cost of accepting the hypothesis that the drug is effective when in fact it is not, or of failing to accept the hypothesis that it is effective when in fact it is, may involve heavy costs in suffering and lives in the case of the cancer drug, but only minor inconvenience in the case of hangnail treatment. For the class of investigations known as academic, which by definition have no tangible value, the cost of a Type I error typically is having an incorrect view of the world, and of a Type II error it is failure to have a correct view. These costs are not easily quantified, but it is essential that some prior estimate be made. However, it is important to realize that if the costs cannot be clearly specified, then any rejection criterion is arbitrary.

The life and social sciences have almost universally accepted a rejection criterion of 5%. That is, they accept a 5% risk of failing to reject a false hypothesis. This criterion, it must be noted, now has the force of tradition, not reason, behind it. Without precise costs attached to each type of error a 5% criterion is as good as any. Also note that there is no traditional criterion for a Type II error. In general, for a given sample size, the probability of a Type II error is inversely related to the probability of a Type I error. This group of sciences has adopted the tactic of fixing the probability of a Type I error and then obtaining a sample as large as possible. A large sample will tend to reduce the probability of a Type II error, but not in a simple or obvious way.

At any rate, recognizing the arbitrariness of it, a 5% rejection criterion will be used here.

The Logarithmic Transformation

In the following sections we shall require the use of the logarithmic transformation of some expressions. Only natural logarithms, i.e. to base e $\sim$ 2.7, are used. For those who do not remember, the rules of logarithms are presented below for reference:

Expression	Transformation
e^b	b
$a \cdot b$	ln(a) + ln(b)
a^b	b ln(a)
a/b	ln(a) − ln(b)
a^{-b}	− b ln(a)

Notice that the logarithmic operator is indicated notationally by "ln" for "logarithm natural."

Also the following notational conventions will be used throughout this chapter:

i, j, k : subscripts, indexes
x : observed frequency
μ : expected frequency
m : estimate of expected frequency
π : probability
p : estimate of probability (proportion)

5.2. Testing a One-Dimensional Hypothesis

A one-dimensional hypothesis asserts that all categories are equally probable. For example, the one-dimensional hypothesis about the touching behavior of the deBraza monkeys described in the last chapter is that all 6 events—BC, BD, CB, CD, DB, DC—are equally likely; each event has probability 1/6. If primate mating systems are polygyny, polyandry, promiscuity, and monogamy the one-dimensional hypothesis asserts that each has probability 1/4.

Let the frequency of category i be $x_i : i = 1, 2, \ldots, k$ for a k category variable. By hypothesis, the estimate of the expected frequency in each category is

$$m_i = (1/k) \sum_i x_i. \tag{5.1}$$

Also by hypothesis, the ratio (x_i/m_i) should be 1.0.

The log-likelihood ratio chi-square statistic, defined by

$$G^2 = 2 \sum_i x_i(\ln(x_i/m_i)) \tag{5.2}$$

has approximately a chi-square distribution with k-1 degrees of freedom (Haberman, 1978; Fienberg, 1977, 1980). Notice that if the hypothesis is true, $G^2 = 0$ because $\ln(1) = 0$.

There are some things to note about equations 5.1 and 5.2. Clearly an $m_i = 0$ will create problems in 5.2. This would happen if $\sum_i x_i = 0$, that is when an observed frequency is zero. In the case at hand this would be more nuisance

Table 5.2.1. Touching Events among Captive deBraza Monkeys

Event	Frequency	Expected	Ratio	$x_i \ln(x_i/m_i)$
BC	8	15.3	0.523	−5.187
BD	9	15.3	0.588	−4.776
CB	6	15.3	0.392	−5.617
CD	7	15.3	0.458	−5.474
DB	46	15.3	3.007	50.636
DC	16	15.3	1.046	0.716
Total	92			30.298

$G^2 = 2(30.298) = 60.6$, df = 5.

than crisis as the category would simply be eliminated. In general, however the matter is more troublesome. Consult Fienberg (1977, 1980) or Bishop, Fienberg, and Holland (1976) for guidance. Secondly, note that m_i is the estimate of the expected value under the hypothesis of equal probabilities for all k categories. G^2 is defined in terms of the ratio of observed to expected frequency.

Consider the deBraza monkeys (Sec. 4.3.3). Recall that the notation BC means B touches C. The data are in Table 5.2.1. By hypothesis, all events are equally likely so the expected frequency is a constant: expected frequency = $Np = 92(1/6) = 15.3$. The ratio of observed to expected is obtained in the obvious way, e.g. for the event BC the ratio is $8/15.3 = 0.523$. The log of the ratio, e.g. for event BC, $\ln(0.523) = -0.648$, is then multiplied by the observed frequency, to obtain the last column, e.g. $(-0.648)(8) = -5.184$. Note that rounding error accounts for the difference between -5.194 and the totaled value -5.187. Finally the entries in the last column are summed, and multiplied by 2 to produce $G^2 = 60.6$.

Relative frequencies are displayed in Figure 5.1.1. It is always a good idea to produce some kind of graphic for data. The impact of a picture is, for most of us, greater than a column of numbers. With $6 - 1 = 5$ degrees of freedom, a value of G^2 greater than or equal to 11.1 allows rejection of the hypothesis of equal probabilities such that the probability of being wrong—committing a Type I error—is less than 0.05. (An attenuated table of the chi-square distribution will be found in Appendix D.) Note that this probability is directly under experimenter control. Do not be tempted into making this probability very small just to exercise your authority. The reason is that as the probability of a Type I gets small, the probability of a Type II gets large for a given sample size. This means that you should give considerable thought to the selection of this probability. Specifically it should be chosen so as simultaneously to minimize the cost of (1) deciding that a false hypothesis is true, and (2) deciding that a true hypothesis is false. This sounds simple. In fact it is not, either mathematically or empirically. The difficulty has produced a curious reaction

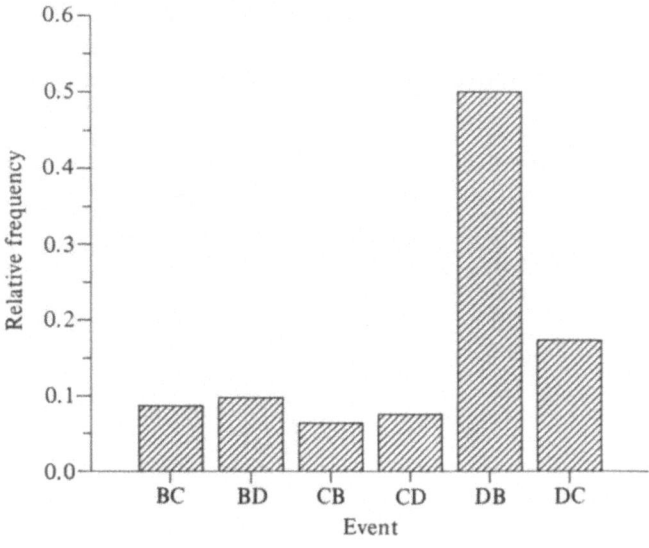

Figure 5.1.1

—adherence to traditional practice. At best, this is an abrogation of responsibility. Part of the difficulty stems from the fact that determining the appropriate costs is troublesome and more so in academic fields such as anthropology. It is relatively straightforward in applied fields such as medicine or engineering. All the anthropologist risks, after all, is notoriety if he is wrong. And people have short memories for such things. I shall, in conformity with long tradition, wave my hands at the problem and use the 0.05 level. This is for convenience only.

Standardized Residuals

Having observed $G^2 = 60.6$ and rejected the hypothesis, the next step is to examine the deviations from expectation. This is accomplished by evaluating the standardized residuals. A residual is a deviation from expectation. These are standardized by dividing each by an estimate of variability.

The quantity

$$c = Np(1 - p) = 92(1/6)(5/6) = 12.8$$

estimates the variability for these data. We shall actually use the square root, 3.6, as the denominator. The expression

$$R_i = (x_i - m_i)/\sqrt{c} \tag{5.3}$$

is the standardized residual. These are evaluated in Table 5.2.2.

Table 5.2.2. Standarized Residuals
for the Touching Events by the
Captive deBraza Monkeys

Event	Residual	Standardized residual
BC	−7.3	−2.04
BD	−6.3	−1.76
CB	−9.3	−2.60
CD	−8.3	−2.32
DB	30.7	8.59
DC	0.7	0.20

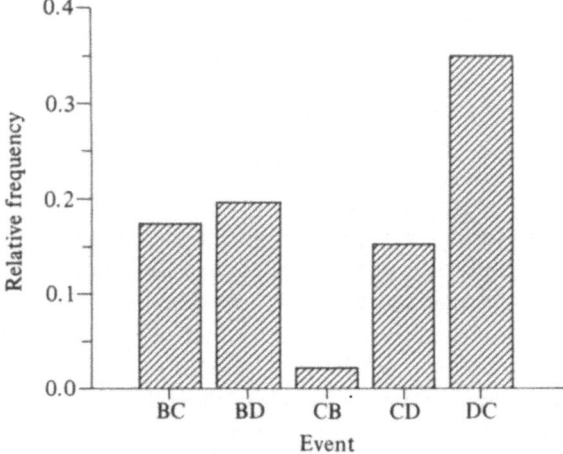

Figure 5.1.2

Table 5.2.3. Touching Events among deBraza Monkeys
Excluding Those Initiated by the Juvenile Toward
Its Mother

Event	Frequency	Expected	Ratio	$x_i \ln(x_i/m_i)$
BC	8	9.2	0.87	−1.11
BD	9	9.2	0.98	−0.18
CB	6	9.2	0.65	−2.59
CD	7	9.2	0.76	−1.92
DC	16	9.2	1.74	8.86
Total	46			3.06

$G^2 = 2(3.06) = 6.12$, df = 4.

Table 5.2.4. Touching Events among deBraza Monkeys
Excluding All Those Initiated by the Youngster

Event	Frequency	Expected	Ratio	$x_i \ln(x_i/m_i)$
BC	8	7.5	1.07	0.54
BD	9	7.5	1.20	1.64
CB	6	7.5	0.80	-1.34
CD	7	7.5	0.93	-0.51
Total	30			0.33

$G^2 = 2(0.33) = 0.66$, df $= 3$.

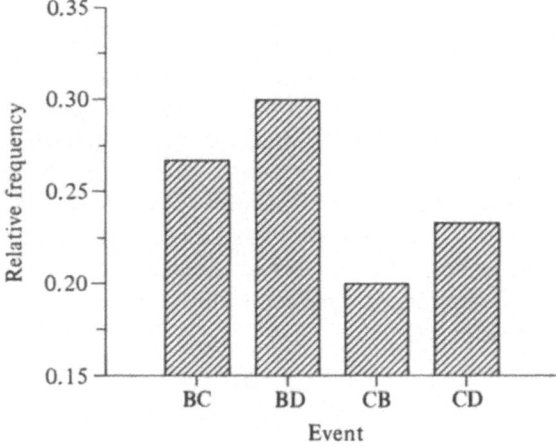

Figure 5.1.3

The expected value of the R_i is 0 because $x_i = m_i$ when the hypothesis is true.

Clearly the main source of the large value of G^2 is the large deviation for the event DB. Recall that B is the mother of D.

It might be instructive to eliminate the DB events and re-evaluate the hypothesis. Note that this is an exploratory process and is not a proper test of a hypothesis. The goal is to elaborate our understanding of the social organization of the monkeys. Any conclusions would necessarily be tentative and subject to independent confirmation. Consider Fig. 5.1.2 and Table 5.2.3. Note that even though the value of G^2 is less than the 5% criterion of 9.49, the activity initiated by the young animal still seems to distort the structure. If we now eliminate the events initiated by the youngster and directed toward the other female, Table 5.2.4 and Figure 5.1.3 result. It appears that the fit to the hypothesis of random events has been improved considerably by eliminating the events initiated by the youngster. This suggests that these touching

Table 5.2.5. Marital Residence Patterns in the World
Ethnographic Sample

Pattern	Frequency	Expected	Ratio	$x_i \ln(x_i/m_i)$
Patrilocal	154	47.8	3.222	180.2
Bilocal	12	47.8	0.251	-16.6
Matrilocal	51	47.8	1.067	3.3
Avunculocal	18	47.8	0.377	-17.6
Duolocal	4	47.8	0.084	-9.9
Total	239			139.4

$G^2 = 2(139.4) = 278.8$, df $= 4$.

events have two distinct structural components: the set of events initiated by the youngster, and the set initiated by anyone else. The former is highly structured and the latter is apparently random. (The adult male constitutes a third component of the social structure which has been ignored.)

The World Ethnographic Sample (Murdock, 1967) categories and frequencies of marital residence pattern are presented in Table 5.2.5. The estimate of the expected frequency of each pattern is $m_i = 239(1/5) = 47.8$. The value of $G^2 = 278.8$ is referred to a table of the chi-square distribution with 4 degrees of freedom. The critical 5% value is 9.49 so the (one-dimensional) hypothesis of equal frequencies is confidently rejected. The denominator for determining the standardized residuals is

$$\sqrt{c} = (239(1/5)(4/5))^{1/2} = 6.18.$$

5.3. Testing a Two-Dimensional Hypothesis

5.3.1. The 2×2 Table

5.3.1.1. BLOODGROUPS AND ILLNESS

With two dimensions, two variables, the number of possible hypotheses is more than doubled. Consider an example. Woolf (1955) obtained a very large number of observations of blood group phenotypes A and O for people with alimentary ulcers and a control group. The observations were collected in three cities of the U.K., but we shall ignore the city variable. The results are presented in Table 5.3.1 and the odds for phenotype O relative to A are displayed in Figure 5.2. Three questions may be asked about such data:

(1) Are the frequencies of the blood group types among patients with ulcers different from those among controls?
(2) Is the frequency of ulcers different among people with type O from that among people with type A?
(3) Are the variables, blood group type and ulcers, independent?

Table 5.3.1. The Joint Distribution of
Ulcers and Blood Group Phenotypes
A and O

	Blood group type		
Condition	O	A	Total
Ulcer	1668	1044	2712
Control	15708	13255	28963
Total	17376	14299	31675

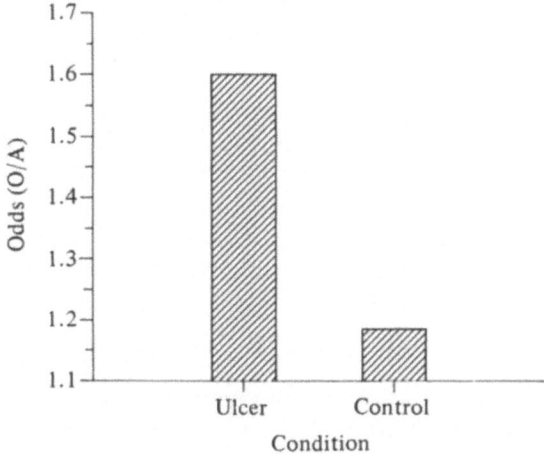

Figure 5.2

As it turns out these are all variants of the same question. But they do have
distinctly different origins in the design of the research. If the project is designed to make exactly 31675 observations, it is called a fixed total design.
If the total number of cases of ulcers is fixed at 2712 and controls at 28963,
the row margins are fixed. If the number of cases of type O is fixed at 17391
and of type A at 14299 the column marginals are fixed.

In order to proceed with the analysis, the following notational conventions
will be used.

	Blood group type		
	O	A	Total
Ulcer	x_{11}	x_{12}	x_{1+}
Control	x_{21}	x_{22}	x_{2+}
Total	x_{+1}	x_{+2}	x_{++}

Table 5.3.2. Evaluation of the Expected Values under the Hypothesis of
Independence for Table 5.3.1

	Marginal	Frequency	ln(Frequency)
	1+	2712	7.91
	2+	28963	10.27
	+1	17391	9.76
	+2	14299	9.57
	++	31675	10.36

Condition	Blood group type	$\ln(x_{i+})$	$\ln(x_{+j})$	$\ln(x_{++})$	$\ln(m_{ij})$	m_{ij}
Ulcer	O	7.91	9.76	10.36	7.31	1487.7
	A	7.91	9.57	10.36	7.11	1222.9
Control	O	10.27	9.76	10.36	9.67	15898.8
	A	10.27	9.57	10.36	9.48	13072.2

First consider the case with the total, x_{++}, fixed. The cell probabilities are
estimated by the proportions

$$p_{ij} = x_{ij}/x_{++}. \tag{5.4}$$

From results in the previous chapter we may observe that the expected frequencies, under the hypothesis of independence, are given by

$$m_{ij} = (x_{++})(p_{i+})(p_{+i}) \tag{5.5}$$

where $p_{i+} = x_{i+}/x_{++}$. This may be written as

$$m_{ij} = (x_{i+})(x_{+j})/x_{++}.$$

Now taking (natural) logarithms, equation 5.5 becomes

$$\ln(m_{ij}) = \ln(x_{i+}) + \ln(x_{+j}) - \ln(x_{++}). \tag{5.6}$$

Notice that equation 5.6 is linear—it involves only addition and subtraction
—in the logarithms of the table marginals. It is called a log-linear expression
because it is linear in the logarithms. The equation is evaluated for the example
in Table 5.3.2. (This would be a good place to confirm that you are able to
obtain natural logs and their inverse. For example, find the natural log of 3,
$\ln(3) = 1.10$. Now find $\exp(1.10) = \ln^{-1}(1.10) = 3$.)

Comparing the expected values from the right-most column of Table 5.3.2.
with the observed values, there are some apparently large discrepancies. For
example, if condition is independent of blood group type, then one expects
1487.7 cases of ulcers and type O blood; the observed frequency is 1668. From
these discrepancies it seems that type O may dispose one to acquire ulcers.

Table 5.3.3. Evaluation of Log-Likelihood Ratio Chi-Square for Table 5.3.1

Condition	Blood group type	Observed	Expected	$O\ln(O)$	$O\ln(m)$
Ulcer	O	1668	1487.7	12375.5	12184.7
	A	1044	1222.9	7256.7	7421.8
Control	O	15708	15898.8	151769.5	151959.2
	A	13255	13072.2	125818.2	125634.1
Total				297219.9	297199.8

$G^2 = 2[297219.9 - 297199.8]$
   $= 40.2.$

The log-likelihood ratio chi-square statistic is

$$G^2 = 2 \sum (\text{observed}) \ln(\text{observed/expected}) \qquad (5.7)$$

$$= 2[\sum (\text{observed}) \ln(\text{observed}) - \sum (\text{observed}) \ln(\text{expected})] \qquad (5.8)$$

which is evaluated in Table 5.3.3. A 2×2 table has 1 degree of freedom. The critical value of chi-square at the 5% level is 3.84. The obtained value is larger than the critical value so we may reject the hypothesis of independence. Using the 5% criterion has the following interpretation. If the variables are actually independent and if we could obtain a large number of replications of this experiment, then 5% of the calculated G^2 values would be greater than 3.84. Notice that the fact that the computed value is much larger than the criterion does not enter the interpretation. Specifically we might be tempted to entertain the following argument: since the observed G^2 is about 10 times greater than the criterion, the true significance level of the test is about 0.5%, i.e. we have 99.5% confidence in rejecting the hypothesis. This is a fallacy.

5.3.1.2. AGNATIC AND UTERINE ALTRUISM

Agnatic relatives are related through male geneological links and uterine relatives through female links. With regard to conjugal stability and geneological souce of altruism, Flinn (1981) observes:

> "... altruism dispensed by a man to his offspring might be utilized by his offspring's uterine half-siblings (or other relatives of his ex-wife) who are unrelated to him... This effect of divorce is similar to paternity uncertainty." (*ibid.*, 448)

The expectation is that conjugal stability will be high where agnatic kin are the primary source of altruism. Conversely, where uterine kin are the

Table 5.3.4. Conjugal Instability and
Primary Source of Altruism

Conjugal instability	Source of altruism		Total
	Agnatic	Uterine	
Low	94	0	94
High	14	18	32
Total	108	18	126

Table 5.3.5. Standardized
Residuals for the Survey of
Conjugal Instability and
Primary Source of Altruism

Conjugal instability	Source of altrusim	
	Agnatic	Uterine
Low	1.5	−3.7
High	−2.6	6.3

primary source of altruism, conjugal stability will be low. The World Ethnographic Sample and other sources were surveyed for information on conjugal instability and primary source of altruism. Partial results are in Table 5.3.4. You should confirm the computed value of $G^2 = 59.5$. With one degree of freedom, this allows rejection of the default null hypothesis of no relationship. In Table 5.3.5. are the estimated standardized residuals. Note that the relationship is as predicted, i.e. a deficit of cases with (1) an agnatic source of altruism and high frequency of conjugal instability, and (2) a uterine source and low frequency of instability. Excess cases are observed on the other diagonal. The observed odds are displayed in Figure 5.3.

5.3.1.3. SUICIDE AMONG AMERICAN PHYSICIANS, THE FIRST STUDY

The study of the frequency of suicide among American physicians that was mentioned earlier has a three-dimensional structure. In a later section (Sec. 5.4.1) we shall treat all dimensions simultaneously. In this section the 3-way table is decomposed into all possible 2-way tables. You should note that the sampling on this particular project was as follows: the focus of attention was on psychiatrists and so all obituaries for psychiatrists were observed; only about ten percent of the obituaries of other specialties was observed. This

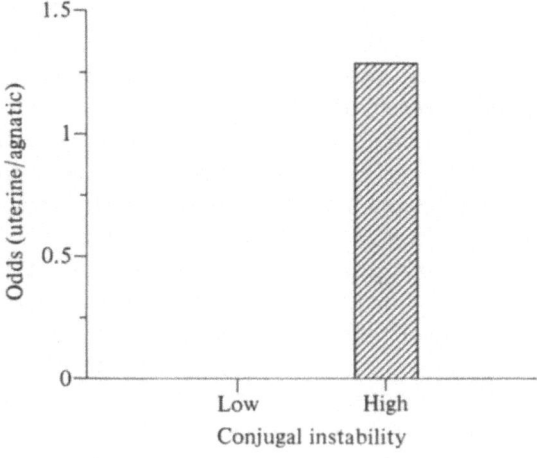

Figure 5.3

Table 5.3.6. Death Types by Age and Specialty
among American Physicians

| | | Age | | |
		25–54	55–79	Total
Non-psychiatry	Suicide	205	148	353
	Other	352	520	872
	Total	557	668	1225
Psychiatry	Suicide	42	9	51
	Other	21	30	51
	Total	63	39	102

means that the project shall pay no attention to determining the true rate of
suicide in the population of all physicians. But questions can be asked about
differences between types of physicians and/or among the age groups. The
data are presented in Table 5.3.6. Suppose that we have the research hypothesis that suicide rate increases with age. This is easily rationalized with
reference to the epidemiological effect of "time at risk." This hypothesis can
be formulated so that it predicts that the rate of suicide is less in the age group
under 55 than in the age group over 55. In the context of this project, what
are we to make of the assertion [*not* P]? Clearly [*not* P] means that the suicide
rate in the older age group is not greater than the rate in the younger age
group, that is, the old rate is less than or equal to the young rate.

Death Type by Age

From Table 5.3.6 we may estimate that the probability of suicide in the younger age group is given by $P_1(S) = (205 + 42)/(557 + 63) = 0.398$. In the older age group we get that $P_2(S) = (148 + 9)/(668 + 39) = 0.222$. Since the probability in the older age group is smaller than that in the younger age group we may justifiably conclude at this point that the frequency of suicide does not increase with age. This conclusion is possible only because the prediction was quite specific about the direction of the effect. If the prediction had been nondirectional, "there is a difference between young and old MDs with regard to suicide rate," then we could not reach any conclusion on the basis of the observations so far. As a rule, if the theory is too vague to be specific about the direction of an effect, it is usually not worthwhile to obtain data. The theory should be refined prior to further work.

A general procedure for the determination of the truth value of the prediction will now be presented. The null hypothesis to be tested, specifically the negation of the consequent, asserts that these two variables, age and death type, are not positively correlated. If the variables are independent then they are neither positively nor negatively correlated. So if independence is accepted, the null hypothesis is accepted. If independence is not accepted, further considerations are required to determine whether the correlation is positive or negative. Assuming, then, that these are two independent binomial variables, we get the following unconditional estimates of the parameters of the two binomial variables:

$$P(\text{suicide}) = 404/1327 = 0.30 \qquad P(\text{young}) = 620/1327 = 0.47$$

$$P(\text{other}) \ = 923/1327 = 0.70 \qquad P(\text{old}) \quad = 707/1327 = 0.53.$$

Note that in obtaining these it has been assumed that the sample size was fixed at 1327. Under the hypothesis of independence we expect the following proportions in each of the four categories of the conjunction of these two variables:

	Age	
	25–54	55–79
Suicide	$(0.30) \cdot (0.47)$ $= 0.14$	$(0.30) \cdot (0.53)$ $= 0.16$
Other	$(0.70) \cdot (0.47)$ $= 0.33$	$(0.70) \cdot (0.53)$ $= 0.37$

These are the expected proportions under the hypothesis of independence. They allow us to obtain the expected values.

Expected Values for Age by Death Type
Under the Hypothesis of Independence

	Age	
	25–54	55–79
Suicide	(0.14)·1327 = 188.4	(0.16)·1327 = 216.3
Other	(0.33)·1327 = 431.3	(0.37)·1327 = 491.0

It is the agreement between the observed values and the expected values which allow a decision to be made regarding the hypothesis being tested. Specifically the quantity to be formed is given by equation 5.8. The observed frequencies are presented in Table 5.3.7. The computation of G^2 for the data in Table 5.3.7 is illustrated in Tables 5.3.8 and 5.3.9. In Table 5.3.8 the expected values are obtained according to equation 5.6. And in Table 5.3.9 is the evaluation of G^2.

A concept which causes considerable conceptual difficulty for beginning students is that of degrees of freedom. The following discussion is intended to motivate your intuition. The complete explanation is far beyond the scope of this so eventually you will be asked to accept a statement. Consider a binomial variable. There are two parameters, Π and $1 - \Pi$, that characterize this distribution. However these parameters are not independent. If one is estimated then the other is known directly and immediately. Consequently there is a single degree of freedom for a binomial variate. In the multinomial case with n categories the sum of the probabilities still must be identically equal to 1, there is some greater latitude, "degrees of freedom," regarding $n - 1$ of the values. But once the $n - 1$ values are specified then the nth value is given directly. Consequently there are $n - 1$ degrees of freedom for this particular multinomial. The degrees of freedom for the conjunction of binomial or multinomial variates is simply the product of the degrees of freedom for each

Table 5.3.7. Death Types by Age
among American Physicians

Death type	Age		Total
	25–54	55–79	
Suicide	247	157	404
Other	373	550	923
Total	620	707	1327

$G^2 = 48.7$, df = 1.

Table 5.3.8. ln Expected Values for Death Type by Age

	Marginals	Frequency	ln(Frequency)
	1+	404	6.00
	2+	923	6.83
	+1	620	6.43
	+2	707	6.56
	++	1327	7.19

Death type	Age	$\ln(x_{++})$	$\ln(x_{i+})$	$\ln(x_{+j})$	$\ln(m_{ij})$	m_{ij}
1	1	7.19	6.00	6.43	5.24	188.7
1	2	7.19	6.00	6.56	5.37	214.9
2	1	7.19	6.83	6.43	6.07	432.7
2	2	7.19	6.83	6.56	6.20	492.7

Table 5.3.9. Evaluation of G^2 for Death Type by Age

Death type	Age	Observed	Expected	$O\ln(O)$	$O\ln(x^*)$
1	1	247	188.7	1360.8	1294.3
1	2	157	214.9	793.8	843.1
2	1	373	432.7	2208.7	2264.1
2	2	550	492.7	3470.5	3410.0
Total				7833.8	7811.5

$G^2 = 2[7833.8 - 7811.5] = 44.7$.

of the component distributions. Specifically note that Table 5.3.7 constitutes the conjunction of two binomial variates, age and death type. Consequently this table has $(2 - 1) \cdot (2 - 1) = 1$ degrees of freedom. Notice that this is a property of the table, not the sample size. It is necessary to know the degrees of freedom for a table in order that the probability of G^2 may be determined. To determine the probability of G^2 one simply consults a table of chi-square values, entering the table with the degrees of freedom for the table producing the value of G^2. An attenuated table of the values of chi-square by degrees of freedom is available in the Appendix D. The tabled value for 1 degree of freedom at the 5% level is 3.84. Since the computed value is greater than this, we reject the null hypothesis and decide that the variables are not independent.

It is also useful to have a means of comparing the observed and expected frequencies. We know from the value of G^2 that the two variables are not independent of each other. And we suspect, based on the observation that the difference between the relative frequency for (young-MD) minus (old-MD) is

Table 5.3.10. Standardized
Deviates of Observed and
Expected Frequencies in
Death Type by Age

	Age	
	25–54	55–79
Suicide	4.2	−4.0
Other	−2.8	2.6

not negative, that the frequency of suicide tends to decline with age. A useful technique for the comparison between observed and expected frequencies is

$$R_{ij} = (\text{Observed-Expected})/\sqrt{(\text{Expected})}. \qquad (5.9)$$

(You may be interested to notice that this measure of deviation, R_{ij}, is simply the (signed) square root of the contribution of a particular cell, the *ij* cell, to the value of the Pearson chi-square.)

The computed values for this measure of residual variation are presented in Table 5.3.10. Notice that the measure is large for suicide among young MDs and small for suicide among old MDs, i.e. the observed frequency of suicide among young MDs is greater than expected while the observed frequency for suicide among the older MDs is less than expected. This is sufficient to establish that the correlation is not positive. We have accepted the negation of the research hypothesis—suicide rate does not increase with age.

Medical Specialty by Age

We may suspect that the previous result is due to an age difference between psychiatry and other MDs. Notice that we have arrived at a point of knowing one thing that suicide is not. It is not positively related to age. (In separating the population into two different populations, psychiatrists and non-psychiatrists, we have begun a process of exploration in an attempt to learn something positive about the nature of suicide. The outcome of this *ad hoc* process cannot result in a scientifically valid conclusion about the nature of suicide. At the best it may result in an elaborated theory and a new research hypothesis, the evaluation of which would require another project.) The observed frequencies are presented in Table 5.3.11. Notice that there are about twelve times as many non-psychiatrists as there are psychiatrists in this sample of observations. This should indicate to you why the measure of deviations from expectations that we considered above must be standardized in some way. Due to the very large difference in sample sizes, a fractionally small deviation from expectation among non-psychiatrists would appear as an

Table 5.3.11. Specialty by Age among
American Physicians

	Age		
	25–54	55–79	Total
Non-psychiatry	557	668	1225
Psychiatry	63	39	102
Total	620	707	1327

$G^2 = 10.1$, df $= 1$.

Table 5.3.12. ln Expected Values for Specialty by Age

		Marginals		ln
		$1+$	$= 1225$	7.11
		$2+$	$= 102$	4.63
		$+1$	$= 620$	6.43
		$+2$	$= 707$	6.56
		$++$	$= 1327$	7.19

Specialty	Age	$\ln(x_{++})$	$\ln(x_{i+})$	$\ln(x_{+j})$	$\ln(m_{ij})$	m_{ij}
1	1	7.19	7.11	6.43	6.35	572.5
1	2	7.19	7.11	6.56	6.48	652.0
2	1	7.19	4.63	6.43	3.87	47.9
2	2	7.19	4.63	6.56	4.00	54.6

Table 5.3.13. Evaluation of G^2 for Specialty by Age

Specialty	Age	Observed	Expected	$O\ln(O)$	$O\ln(m)$
1	1	557	572.5	3521.7	3537.0
1	2	668	652.0	4344.9	4328.7
2	1	63	47.9	261.0	243.8
2	2	39	54.6	142.9	156.0
Total				8270.5	8265.5

$G^2 = 2[8270.5 - 8265.5] = 10.0$, df $= 1$.

absolutely large deviation relative to that observed in the smaller sample. So some standardization is required.

You should satisfy yourself that the value of G^2 for this table is 10.1, with 1 degree of freedom. The computation of the expected values is illustrated in Table 5.3.12 and G^2 is evaluated in Table 5.3.13. This value of G^2 is improbable if the null hypothesis of independence of age and medical specialty is true.

Table 5.3.14. Standardized
Residuals in Table 5.3.11

	Age	
	25–54	55–79
Non-psychiatry	−0.6	0.6
Psychiatry	2.2	−2.1

Table 5.3.15. Specialty by Death Type
among American Physicians

	Death type		
	Suicide	Other	Total
Non-psychiatry	353	872	1225
Psychiatry	51	51	102
Total	404	923	1327

$G^2 = 18.5$, df $= 1$.

Now consider Table 5.3.14 which contains the standardized residuals. There you will note that the deviation from expectation for young psychiatrists is large and the deviation for old psychiatrists is small. There is an excess of young psychiatrists. This is troublesome. It suggests the possibility that the inflated frequency of suicide among young MDs in Table 5.3.7 may be due to the excess of psychiatrists among young MDs.

Medical Specialty by Death Type

In an attempt to remove this "confounding" of effects, we might check for the suicide frequency among the medical specialties. If it should be the case that psychiatrists are less apt to suicide than are other medical specialties then we might be more confident of the generality of the age effect obtained from Table 5.3.7. The necessary observations are presented in Table 5.3.15. The value of G^2 for this table is 18.5, with 1 degree of freedom. This is an improbable result if the suicide rate is the same among psychiatrists and non-psychiatrists. In Table 5.3.16 are presented the deviations of the observed and expected frequencies. You will note that there is an excess of suicides among psychiatrists and a deficit among non-psychiatrists. So the death type distributions of the two populations cannot be considered to be the same.

Now let us review what we have done so far. In Table 5.3.7 it was observed that death type is not independent of age. In Table 5.3.11 it was observed that medical specialty is similarly not independent of age and consequently it is

Table 5.3.16. Standardized
Residuals in Table 5.3.15

| | Death type | |
	Suicide	Other
Non-psychiatry	-1.0	0.7
Psychiatry	3.6	-2.4

possible that the apparent dependence of suicide frequency on age is nothing more than an artifact of the differences among the medical specialties with regard to age. In Table 5.3.15 it was concluded that the medical specialties are indeed different with regard to the frequency of suicide.

Notice that this kind of procedure, specifically that of constructing all possible two-way tables, obscures two potentially very important facts about frequencies. First by considering only marginal distributions in a serial fashion such as this it is very difficult to make any sense out of the result. For example when we concluded that suicide is not independent of age and then next concluded that the medical specialties are also not independent of age we are then led to suspect that the conclusion in Table 5.3.15 that medical specialty is not independent of death type may be a spurious result. That is, the conclusion from Table 5.3.7 may be the spurious result of the frequencies in Tables 5.3.11 and 5.3.15. Second, there is no way of determining whether or not all three variables need to be considered simultaneously. That is to say the frequency of suicide may in fact depend not just on age or on medical specialty but on the joint, simultaneous, effect of age and specialty. Also recall that our goal is to evaluate the (null) hypothesis that suicide frequency is not positively correlated with age. Considering this, the relationship between medical specialty and death type as well as the relationship between medical specialty and age should be considered nothing but noise. That is these effects, in so far as they exist, only serve to obscure the primary effect of interest which is specifically the relationship between suicide frequency and age.

5.3.2. The 2 × C Table

5.3.2.1. DISCRETE LEVELS

The logic developed for the 2 × 2 table extends in a natural way to accommodate the 2 × C case with C > 2. The degrees of freedom are C-1. The main reason for treating this situation separately is to consider later the special case of ordered categories of the variable with C levels.

Flinn (1981) also considers cousin marriage preference with regard to primary source of altruism.

"... where the father's brother (or other agnatic kin) are important sources of altruism, father's brother's daughter marriage will be preferred; where mother's brother is an important source of altruism, mother's brother's daughter marriage will be preferred; where the father's sister is an important source of altruism, father's sister's daugher marriage will be preferred ..." (*ibid.*, 456)

As an exercise, let the following variable definitions be given:

K : kin selection theory
A : source of altruism
C : preferred cousin.

The first premise of the argument is

if [kin selection theory]
then [source of altruism will determine (probabalistically) the preferred cousin marriage]

which may be rendered:

if [K] *then* [*if* A *then* C].

Note that if we model the event "*not* (*if* [source of altruism] *then* [preferred cousin])" by the condition of independence

$$P(C|A) = P(C) \cdot P(A),$$

that is, [preferred cousin] is independent of [source of altruism], then we may assert the argument

if [not K] *then* [*not* (*if* A *then* C)]
[*if* A *then* C]
∴ K.

Note that the first premise above is just condition 2 of a good test. Also note that if "preferred cousin marriage" is independent of "source of altruism" when "kin selection theory" is false then the project is believable; otherwise it is not. Partial results are presented in Table 5.3.17. You should confirm that $G^2 =$

Table 5.3.17. Cousin Marriage Preference with Regard to Primary Source of Altruism

| | Source of altruism | | |
Cousin	Agnatic	Uterine	Total
Father's sister's daughter	3	6	9
Mother's brother's daughter	39	5	44
Father's brother's daughter	14	11	14
Total	56	11	67

Table 5.3.18. Standardized Residuals for the Survey of Cousin Marriage Preference and Primary Source of Altruism

	Source of altruism	
Cousin	Agnatic	Uterine
Father's sister's daughter	−1.6	3.7
Mother's brother's daughter	0.4	−0.8
Father's brother's daughter	0.7	−1.5

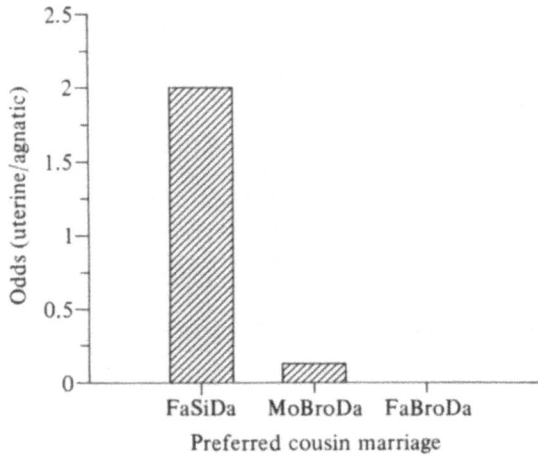

Figure 5.4

17.2 and there are 2 degrees of freedom. In Table 5.3.18 are the standardized residuals. Note that in all cases the direction of the deviation is as predicted. The observed odds are displayed graphically in Figure 5.4.

5.3.2.2. ORDERED LEVELS OF ONE VARIABLE

Alfred (1980) reports frequencies of the sickle-cell trait by age obtained from various screening clinics in the United States. There was a widespread expectation among anthropologists and geneticists that the relative frequency should decline with age. This is obtained from a standard model used by epidemiologists to describe the decay of frequency as a function of the time of exposure. The longer one is at risk, the greater the probability of death. That this argument is based on a fallacy is not important here. We wish to test the hypothesis that the frequency of the sickle-cell trait declines with age. A natural way of doing this is to determine the regression of the proportions on the age

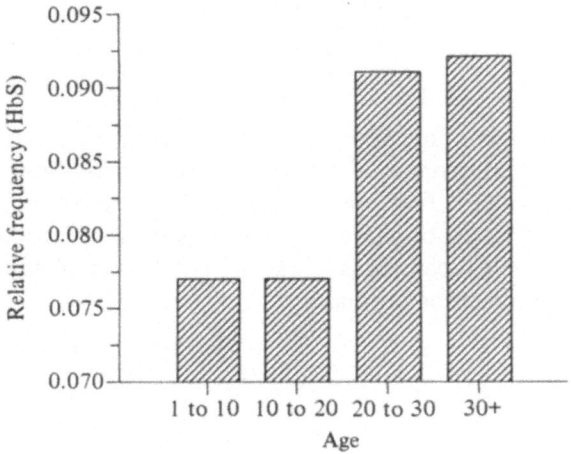

Figure 5.5

Table 5.3.19. Frequency of the Sickle-Cell Trait by Age in the United States

Age category, years	z_j	Normal hemoglobin (n_j)	Sickle-cell trait (x_j)	p_j	$n_j z_j$
1 to <10	−2	12097	1009	0.077	−26212
10 to <20	−1	12065	1008	0.077	−13073
20 to <30	1	3213	320	0.091	3533
>30	2	3257	331	0.092	7176
Total		30632	2668	0.087	−28576
		T	T_x	p	

categories. Then if the regression coefficient is negative we would accept the hypothesis represented by the epidemiological model.

The method presented here is due to Cochran (1954). It uses the "common" chi-square test as distinct from the log-linear version we've been using. The observations are displayed in Figure 5.5 and Table 5.3.19. The column "z_j" contains the numerical scale for the age categories. The n_j are simply the total number of observations in age category j. Then the column "p_j" is obtained by

$$p_j = x_j/n_j.$$

These are just the relative frequency estimates of the probability of sickle-cell trait in age category j. The regression of the p_j on the z_j is the goal. The rightmost column "$n_j z_j$" is required later. Cochran (ibid., 434) defines the weighted regression of the p_j on the z_j as

$$b = \frac{\sum n_j(p_j - p)(z_j - z)}{\sum n_j(z_j - z)^2}$$

where z is the weighted mean of the z_j. Note that the numerator is just the sum of the weighted product of the deviations of the p_j from their mean, p, and the deviations of z_j from their mean, z. This is the weighted covariance of the p_j and z_j. The denominator is the sum of weighted squared deviations of the z_j from their mean. This is the weighted sum of squares. A bit of manipulation of the terms produces

$$\text{Numerator} \quad = \sum x_j z_j - \frac{T_x(\sum n_j z_j)}{T}$$

$$\text{Denominator} = \sum n_j z_j^2 - \frac{(\sum n_j z_j)^2}{T}.$$

Cochran (*ibid.*, 435) then defines the chi-square test for regression, with 1 degree of freedom as

$$\text{Chi-square} = \frac{(\text{Numerator})^2}{pq\,(\text{Denominator})}. \tag{5.10}$$

Note that p with no subscript is the estimated mean and that $q = 1 - p$.
   For the problem of this section we find the following:

$$p = 0.087, \quad q = 0.913, \quad pq = 0.0794$$

$$\text{Numerator} = [(1009)(-2) + (1008)(-1) + (320)(1)$$

$$+ (331)(2)] - (2668(-28576))/30632$$

$$= -2044 + 2488.9$$

$$= 444.9$$

$$\text{Denominator} = [(13106)(-2)^2 + (13073)(-1)^2$$

$$+ (3533)(1)^2 + (3588)(2)^2] - (-28576)^2/30632$$

$$= 49994 - 26658.0$$

$$= 23336.$$

These results produce

$$\text{Chi-square} = \frac{(444.9)^2}{(0.0794)(23336)}$$

$$= \frac{197936}{1852.9}$$

$$= 106.83$$

which is significant at the chosen (5%) level.

Note that knowing that the slope of the line is not zero does not answer the question posed at the beginning. We need to know whether the slope is positive or negative. If it is negative, then the epidemiological hypothesis would be accepted. Otherwise it is rejected. It is not necessary to know the magnitude of the slope, so we can determine by inspection that the slope is not negative. We are therefore able to reject the hypothesis that the relative frequency of sickle-cell trait decreases with age.

5.3.3. The R × C Table

In this section we extend the analytic technique to a two-dimensional table of any size. There are no surprises. A new method of standardizing the residuals will be presented since the variables defining the dimensions are not binomial. More importantly, the results obtained from the example challenge one's commitment to the predetermined level of significance.

Alfred, Grieg, and Petrakis (1979) report on the relationship between a genetic marker, the Duffy system, and educational achievement as measured by highest grade completed. As with many of the other examples presented so far, the argument is the invalid affirm the consequent. This means that the results may be interesting or infuriating but do not compel acceptance.

The observations are presented in Table 5.3.20a.

The cummulative relative frequency of educational achievement by Duffy phenotype is in Figure 5.6.

Note that there are two zero entries in Table 5.3.20a. These are called random zeros and are distinguished from structural zeros. Structural zeros occur with impossible conditions in a table. For example, if age is one dimension and education another, then depending on the categories used, there may be a place in the table for observations of 2 year old Ph.D.'s. As the event is impossible, there can be no observations and a structural zero is produced. A random zero is an artifact of sampling.

The expected values are in Table 5.3.21a.

Table 5.3.20a. Highest School Grade Completed by Duffy Phenotype for Males and Females Older than 25 in Stockton, California

Education	Duffy phenotype				
	Fy(a+,b+)	Fy(a+,b−)	Fy(a−,b+)	Fy(a−,b−)	Total
At most grade 6	3	0	0	6	9
Grade 7 to high school grad.	2	4	11	29	46
More than high school	2	1	3	13	19
Total	7	5	14	48	74

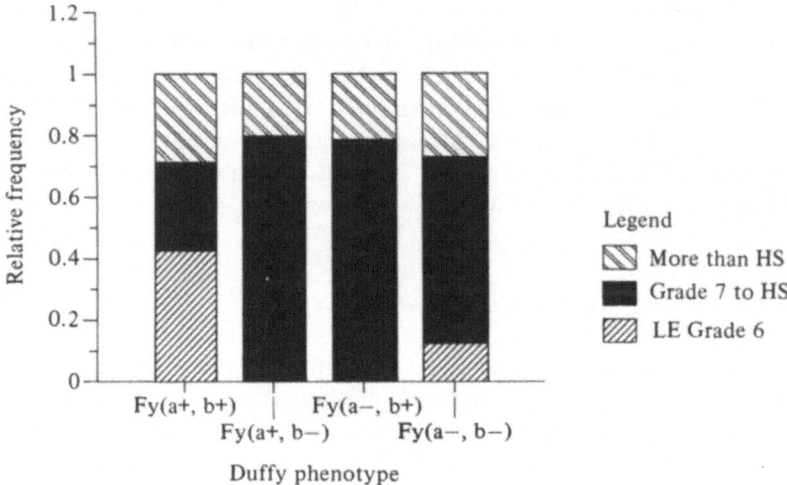

Figure 5.6

Table 5.3.21a. Expected Values for Education by Duffy Phenotype

	Marginals	ln
	$1+\ =9$	2.20
	$2+\ =46$	3.83
	$3+\ =19$	2.94
	$+1=7$	1.95
	$+2=5$	1.61
	$+3=14$	2.64
	$+4=48$	3.87
	$+\ +\ =74$	4.30

Education	Duffy phenotype	$\ln(x_{i+})$	$\ln(x_{+j})$	$\ln(x_{++})$	$\ln(m_{ij})$	m_{ij}
1	1	2.20	1.95	4.30	-0.15	0.86
1	2	2.20	1.61	4.30	-0.49	0.61
1	3	2.20	2.64	4.30	0.54	1.72
1	4	2.20	3.87	4.30	1.77	5.87
2	1	3.83	1.95	4.30	1.48	4.39
2	2	3.83	1.61	4.30	1.14	3.13
2	3	3.83	2.64	4.30	2.17	8.76
2	4	3.83	3.87	4.30	3.40	29.96
3	1	2.94	1.95	4.30	0.59	1.80
3	2	2.94	1.61	4.30	0.25	1.28
3	3	2.94	2.64	4.30	1.28	3.60
3	4	2.94	3.87	4.30	2.51	12.31

Table 5.3.22a. Evaluation of G^2 for Education by Duffy Phenotype

Education	Duffy phenotype	Observed	Expected	$O \ln(O)$	$O \ln(m)$
1	1	3	0.86	3.30	−0.45
1	2	0	0.61	0	0
1	3	0	1.72	0	0
1	4	6	5.87	10.75	10.62
2	1	2	4.39	1.39	2.96
2	2	4	3.13	5.54	4.56
2	3	11	8.76	26.38	23.87
2	4	29	29.96	97.65	98.60
3	1	2	1.80	1.39	1.18
3	2	1	1.28	0	0.25
3	3	3	3.60	3.30	3.84
3	4	13	12.31	33.34	32.63
Total				183.04	178.06

$G^2 = 2(183.04 - 178.06) = 9.98$, df = 6.

Table 5.3.23a. Standardized Residuals for the Education by Duffy Phenotype Data in Table 5.3.20a

Education	Duffy phenotype			
	Fy(a+,b+)	Fy(a+,b−)	Fy(a−,b+)	Fy(a−,b−)
At most grade 6	2.31	−0.78	−1.31	0.05
Grade 7 to high school grad.	−1.14	0.49	0.76	−0.18
More than high school	0.15	−0.25	−0.32	0.20

Notice that these are obtained by equation 5.6. G^2 is evaluated in Table 5.3.22a. The computed probability of a G^2 this large or larger for this table is 0.1041 which, while clearly larger than the 5% criterion is nonetheless disturbingly small. This is a case where a deferred decision is indicated. We would not be justified in rejecting the null hypothesis of independence of Duffy phenotype and educational achievement but the data do suggest the possibility. So rather than abandoning the project we may decide that it is worth pursuing further.

In Table 5.3.23a the standardized residuals are presented. These are calculated according to equation 5.9

$$R_{ij} = (n_{ij} - m_{ij})/\sqrt{m_{ij}} \qquad (5.9)$$

which uses a different denominator than equation 5.3. In ambiguous cases such as this where a decision has been deferred, the reason for calculating and examining the residuals is for guidance with regard to the next step and future research. Note that the (absolutely) large deviations occur in the categories

At most grade 6   *and* $Fy(a+,b+) = 2.31$

At most grade 6   *and* $Fy(a-,b+) = -1.31$

Grade 7 to high
    school grad   *and* $Fy(a+,b+) = -1.14$.

There is no clear pattern to these deviations, so we may try reducing Table 5.3.20a to a 2×2 table using the Duffy phenotypes "$Fy(a+,b+)$" and "Other," and education categories "At most grade 6" and "More than grade 6." This is done in Table 5.3.20b. The expected values are calculated in Table 5.3.21b and G^2 in Table 5.3.22b. (The G^2 value in parenthesis was obtained by computer and is reported so that you may see the effect of rounding error. The probability of a result this large or larger is 0.0281.)

Table 5.3.20b. Reduction of Table 5.3.20a, Highest Grade Completed by Duffy Phenotype in Stockton, California

Education	$Fy(a+,b+)$	Other	Total
At most grade 6	3	6	9
More than grade 6	4	61	65
Total	7	67	74

Table 5.3.21b. Expected Values for Education by Duffy Phenotype, from Table 5.3.20b

	Marginals	ln
	$1+ = 9$	2.20
	$2+ = 65$	4.17
	$+1 = 7$	1.95
	$+2 = 67$	4.21
	$++ = 74$	4.30

Education	Duffy phenotype	$\ln(x_{i+})$	$\ln(x_{+j})$	$\ln(x_{++})$	$\ln(m_{ij})$	m_{ij}
1	1	2.20	1.95	4.30	-0.15	0.86
1	2	4.17	1.95	4.30	1.82	6.17
2	1	2.20	4.21	4.30	2.11	8.25
2	2	4.17	4.21	4.30	4.08	59.15

Table 5.3.22b. Evaluation of G^2 for Education by Duffy Phenotype in Table 5.3.20b

Education	Duffy phenotype	Observed	Expected	$O \ln(O)$	$O \ln(m)$
1	1	3	0.86	3.30	−0.45
1	2	6	6.17	10.75	10.92
2	1	4	8.25	5.55	8.44
2	2	61	59.15	250.76	248.89
Total				270.36	267.80

$G^2 = 2(270.36 - 267.80) = 5.12$, df = 1.

(G^2 estimated by computer = 4.82.)

Table 5.3.23b. Standardized Residuals for the 2 × 2 Education by Duffy Phenotype Data in Table 5.3.20b

Education	Duffy phenotype	
	Fy(a+,b+)	Other
At most grade 6	2.31	−0.07
More than grade 6	−1.48	0.24

There you will note that G^2 exceeds the 5% significance criterion allowing a decision to be reached. Since this is an *ad hoc* result, however, and since the argument structure of the project is invalid, it can only be used as a guide to further research. It would be unacceptable at this point to conclude that the Duffy phenotype affects educational achievement. There is such a suspicion, however. The standardized residuals are in Table 5.3.23b. Note the excess representation in the "Fy(a+,b+) *and* at most grade 6" category and the deficit in the "Fy(a+,b+) more than grade 6" category.

5.4. Tests of Hypotheses in Three or More Dimensions

5.4.1. The 2 × 2 × 2 Table

5.4.1.1. DEATH TYPE, MEDICAL SPECIALTY, AND AGE

Here we shall continue the analysis of death types among MDs begun in Section 5.3. Recall the objections to the analysis developed there, i.e. the decomposition into all possible 2-way tables eliminates information.

Table 5.4.1. Indexing Structure of
Table 5.3.6

Variable	Index	Maximum
Medical specialty	i	I
Death type	j	J
Age	k	K

		Age	
		25–54	55–79
Non-psychiatry	Suicide	111	112
	Other	121	122
Psychiatry	Suicide	211	212
	Other	221	222

We need to develop a different analytic structure in order to correct this defect. In developing this structure we will refer to Table 5.3.6 which is death type by age by MD specialty. We shall refer to the medical specialty by an index, i, and the death types by an index, j, and the age categories by an index, k, so that any cell in the table may be referenced by x_{ijk}.

Before considering the analysis of Table 5.3.6 it is important that you be able to read the indexing structure of complex tables. It is no longer possible simply to let one index be used for rows and another for columns. The indexing structure of Table 5.3.6 is presented in Table 5.4.1. Notice that there are $I \cdot J \cdot K$, cells in the table. The two-way marginals for medical specialty by death type are formed by $x_{ij+} = x_{ij1} + x_{ij2}$. The two-way marginals for medical specialty by age are formed by $x_{i+k} = x_{i1k} + x_{i2k}$. And the two-way marginals for death type by age are formed by $x_{+jk} = x_{1jk} + x_{2jk}$. The one-way marginals are formed similarly. Be aware that in each instance of a summation it is assumed that the sum proceeds over all values of the indexes which are replaced by plus signs.

It should be intuitively apparent that Table 5.3.6 has considerably greater informational content than is present in all the 2-way tables. In terms of the number of hypotheses which can be evaluated from a three-way table by comparison with the number in a two-way table we shall soon be able to substantiate intuition.

Notice that when the subscript i has the value 1 it refers to non-psychiatry and when its value is 2 it refers to psychiatry. When the subscript j has the value 1 it refers to suicide, and when it has the value 2 the reference is to non-suicides. Similarly when the subscript k has the value 1 it refers to the age category 25–54, and the value 2 refers to the age category 55–79.

There is one more fact which is required in order to develop our analytic structure. The sum of the differences between a series of numbers and their expected value is always identically equal to zero. For example, consider that we have a random variable a_i. Then its expected value is $E(a) = \mu$. Then $(a_1 - \mu) + (a_2 - \mu) + \cdots + (a_n - \mu) = 0$.

Under the hypothesis of the independence of the variables in Table 5.3.6 we obtain the cell probabilities by

$$\Pi_{ijk} = \Pi_i \Pi_j \Pi_k.$$

The right-hand side is estimated by

$$\Pi_i \Pi_j \Pi_k = p_i p_j p_k$$

which allows us to obtain the estimates of the expected cell frequencies by

$$m_{ijk} = p_i p_j p_k N \tag{5.11}$$

where N is the total number of observations in the table. Since

$$p_i = x_{i++}/N$$

$$p_j = x_{+j+}/N$$

$$p_k = x_{++k}/N$$

we may write

$$\ln(m_{ijk}) = \ln(x_{i++}/N) + \ln(x_{+j+}/N) + \ln(x_{++k}/N) + \ln(N)$$

$$= -2\ln(N) + \ln(x_{i++}) + \ln(x_{+j+}) + \ln(x_{++k}). \tag{5.12}$$

You should notice that this is simply a straightforward application of two of the laws of logarithms which were presented earlier.

Now consider that the basic problem is to develop a model which accounts for, that is, adequately describes, the observations which have been obtained. This shall mean for us that we seek a mathematical expression which produces expected values close to the observed values. In the case at hand, we shall determine whether the model of independence produces expected values which are close to the observed values or not. The odds for suicide are displayed in Figure 5.7 by age and specialty.

It is possible that all entries in the table simply represent random fluctuations around a constant, the grand mean. So the first component of the model is this constant, u. We need another term for the effect of the variable which we are calling death type. We shall use the notation $u_{1(i)}$ for this term. Similarly we shall require a term for the effect of age and use the notation $u_{2(j)}$. Lastly, a term is needed for the effect of medical specialty, $u_{3(k)}$.

5.4.1.1.1. The Model of Mutual Independence

Now we are able to write an expression for the expected values in the table as the sum of several different independent effects:

$$\ln(\mu_{ijk}) = u + u_{1(i)} + u_{2(j)} + u_{3(k)}. \tag{5.13}$$

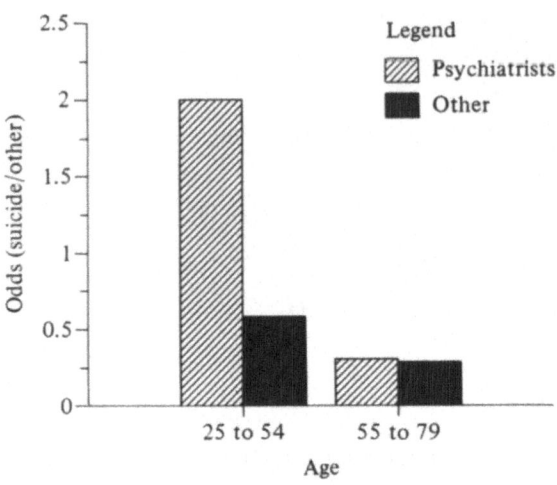

Figure 5.7

The u term with no subscript is the grand mean, a constant. In the new notational convention this is the expression of the model for the expected values expressed in terms of independent effects only.

Still this model is not usable until estimates for the terms are defined. It is quite natural to express the estimate of the grand mean (u with no subscript) as the average of the logs over all observations in the table. Let us also adopt the notational convention that I is the number of rows in the table, J is the number of columns, and K is the number of slabs. In the case at hand $I = J = K = 2$. Then by summing all of the $\ln(x_{ijk})$ and dividing by the number of cells in the table we shall obtain a grand mean of the logs for the table. If the estimate of the row effect is defined as the sum of the row deviations from the grand mean, then the effect of row i may be expressed as the deviation of the marginal for row i from the grand mean. Similarly we can express the estimated effect for column j as the deviation of the marginal for column j from the grand mean. The notation for these concepts is presented as

$$u = \frac{1}{IJK} \sum_i \sum_j \sum_k \ln(x_{ijk}) \tag{5.14a}$$

$$u + u_{1(i)} = \frac{1}{JK} \sum_j \sum_k \ln(x_{ijk}) \tag{5.14b}$$

$$u + u_{2(j)} = \frac{1}{IK} \sum_i \sum_k \ln(x_{ijk}) \tag{5.14c}$$

$$u + u_{3(k)k} = \frac{1}{IJ} \sum_i \sum_j \ln(x_{ijk}). \tag{5.14d}$$

Note that

$$\sum_i u_{1(i)} = \sum_j u_{2(j)} = \sum_k u_{3(k)} = 0.$$

We may now express the model which was defined in equation 5.13 as

$$\ln(m_{ijk}) = -2\frac{1}{IJK}\sum_i\sum_j\sum_k \ln(x_{ijk}) + \frac{1}{JK}\sum_j\sum_k \ln(x_{ijk})$$

$$+ \frac{1}{IK}\sum_i\sum_k \ln(x_{ijk}) + \frac{1}{IJ}\sum_i\sum_j \ln(x_{ijk}).$$

(5.15)

In Table 5.4.2 are computed the expected values under the hypothesis of mutual independence for each of the eight cells of this table. Notice that there are now three subscripted u terms because there are three variables. And in Table 5.4.3 the goodness of fit of these expected values is evaluated. Notice that $G^2 = 79.6$. Also notice that both G^2 and df are now subscripted.

Table 5.4.2. ln Expected Values for Table 5.3.6 under the Hypothesis of Mutual Independence

MD	Death	Age	$u_{1(i)}$ $\ln(x_{i++})$	$u_{2(j)}$ $\ln(x_{+j+})$	$u_{3(k)}$ $\ln(x_{++k})$	u $2\ln(x_{+++})$	$\ln(m)$	m
1	1	1	7.11	6.00	6.43	14.38	5.16	174.2
1	1	2	7.11	6.00	6.56	14.38	5.29	198.7
1	2	1	7.11	6.83	6.43	14.38	5.99	398.1
1	2	2	7.11	6.83	6.56	14.38	6.12	454.0
2	1	1	4.63	6.00	6.43	14.38	2.68	14.5
2	1	2	4.63	6.00	6.56	14.38	2.81	16.5
2	2	1	4.63	6.83	6.43	14.38	3.51	33.1
2	2	2	4.63	6.83	6.56	14.38	3.64	37.8

Table 5.4.3. Evaluation of G^2 for the Hypothesis of Mutual Independence in Table 5.3.6

MD	Death	Age	Observed	Expected	$O\ln(O)$	$O\ln(m)$
1	1	1	205	174.2	1091.2	1057.8
1	1	2	148	198.7	739.6	783.2
1	2	1	352	398.1	2064.0	2107.3
1	2	2	520	454.0	3252.0	3181.4
2	1	1	42	14.5	157.0	112.3
2	1	2	9	16.5	19.8	25.2
2	2	1	21	33.1	63.9	73.5
2	2	2	30	37.8	102.0	109.0
Total					7489.5	7449.7

$G_1^2 = 2[7489.5 - 7449.7] = 79.6$, $\text{df}_1 = IJK - I - J - K + 2 = 4$.

Table 5.4.4. Standardized Deviates for the
Model of Mutual Independence in
Table 5.3.6

		Age	
		25–54	55–79
Non-psychiatry	Suicide	2.3	−3.6
	Other	−2.3	3.1
Psychiatry	Suicide	7.2	−1.9
	Other	−2.1	−1.3

The expression for the degrees of freedom for the hypothesis of mutual independence of three variables is

$$IJK - I - J - K + 2.$$

In this example $I = J = K = 2$, so there are four degrees of freedom. Referring this value to a table of chi-square we discover that 14.86 is the critical value for the test of an hypothesis at the 0.05, or 5%, level of confidence with 4 degrees of freedom. We therefore conclude that the model of independence does not fit the data.

The standardized residuals for the departures from the model of mutual independence are presented in Table 5.4.4. There it will be noted that the largest departure, excess, is that for suicide among young psychiatrists. Less notable but still quite large is the negative departure, deficit, for suicide among older non-psychiatrists. You should be warned against concluding from these results that young psychiatrists are at greater risk of suicide than are young physicians of other specialties. The evaluation of relative risk can only be meaningfully made when we have obtained a model that fits the data acceptably.

The complete general log-linear model for the three-way table is

$$\ln(\mu_{ijk}) = u + u_{1(i)} + u_{2(j)} + u_{3(k)}$$
$$+ u_{12(ij)} + u_{13(ik)} + u_{23(jk)} \tag{5.16}$$
$$+ u_{123(ijk)}.$$

The model of mutual independence is simply the first four terms on the right-hand side of equation 5.16. Then evaluation of the model of mutual independence is equivalent to setting

$$u_{12(ij)} = u_{13(ik)} = u_{23(jk)} = u_{123(ijk)} = 0.$$

This means that all possible two-way interactions between variables as well as higher order interactions are assumed to be zero. Consequently it should come as no surprise that the model of mutual independence does not fit the data. From the earlier examination of two-way tables it became quite clear that there was considerable pairwise dependence between these variables. The hypothesis of mutual independence which excludes pairwise independence should not fit the data.

You should form the habit of thinking of every term in equation 5.16 as a potential model of the data. Furthermore models may be formulated by combinations of terms in this equation. For example it is conceivable to specify a model as $u + u_{12(ij)}$, or $u_{123(ijk)}$.

This is an appropriate time to mention that all log-linear models have a hierarchical structure. Simply stated this means that if a model is specified as $u_{123(ijk)}$ hierarchical structure means that all possible lower order models involving the subscripts are also included in the model. Specifically an alternate way of expression equation 5.16 is

$$\ln(\mu_{ijk}) = u_{123(ijk)}.$$

Similarly, if the model is expressed as $u_{23(jk)}$ this means that

$$u_{12(ij)} = u_{13(ij)} = u_{123(ijk)} = 0$$

and all other terms in equation 5.16 are, by hypothesis, not equal to zero. If the model is specified as $u_{3(k)}$ this means that

$$u_{1(i)} = u_{2(j)} = u_{12(ij)} = u_{13(ik)} = u_{23(jk)} = u_{123(ijk)} = 0.$$

The order (degree) of a model will be defined as the maximum number of subscripts on any term in the model. So equation 5.16 is called a third degree model. A model specified as '$u_{12(ij)} + u_{13(ik)}$' is called a second degree model. (This is a slight departure from common usage which refers to this as a first degree model. The difference originates in the specification of the degree of the model of mutual independence which is usually said to have order zero. It seems more intuitive to refer to model order in terms of subscripts.)

5.4.1.1.2. Independence of One Variable from the Joint Distribution of the Other Two

Earlier we saw that age and medical specialty are probably dependent on each other. Let us determine whether death type is independent of the joint distribution of age and medical specialty. The model is expressed as

$$\ln(\mu_{ijk}) = u + u_{1(i)} + u_{2(j)} + u_{3(k)} + u_{13(ik)}. \tag{5.17}$$

The closed expression for the estimate of the expected values is

$$m_{ijk} = (x_{+j+})(x_{i+j})/x_{+++} \tag{5.18}$$

which on transformation becomes

$$\ln(m_{ijk}) = \ln(x_{+j+}) + \ln(x_{i+k}) - \ln(x_{+++}) \tag{5.19}$$

Table 5.4.5. Age by Medical
Specialty Marginals (x_{i+k})

	Age	
	25–54	55–79
Non-psychiatry	557	668
Psychiatry	63	39

Table 5.4.6. ln Expected Values for Table 4.1 under Model
Equation 5.19

MD	Death	Age	$u_{2(j)}$ $\ln(x_{+j+})$	$u_{13(ik)}$ $\ln(x_{i+j})$	u $\ln(x_{+++})$	$\ln(m_{ijk})$
1	1	1	6.00	6.32	7.19	5.13
1	1	2	6.00	6.50	7.19	5.31
1	2	1	6.83	6.32	7.19	5.96
1	2	2	6.83	6.50	7.19	6.14
2	1	1	6.00	4.14	7.19	2.95
2	1	2	6.00	3.66	7.19	2.47
2	2	1	6.83	4.14	7.19	3.78
2	2	2	6.83	3.66	7.19	3.30

where it is assumed that

$$u_{12(ij)} = u_{23(jk)} = u_{123(ijk)} = 0.$$

Notice that this model description illustrates the hierarchical structure of log-linear models on frequency data. Specifically when a term is included in the model all lower order terms involving the same subscripts are automatically included. Also you should be aware that in general there are three models with this structure.

In order to evaluate the goodness of fit for the expected values to the observations we must first construct the two-way marginals of age and medical specialty. These marginals are presented in Table 5.4.5. The logs of the expected values obtained from equation 5.19 are presented in Table 5.4.6. And these expected values are tested for goodness of fit to the observed frequencies in Table 5.4.7. Notice that this test is computed according to equation 5.19. There are $[(I - 1)(JK - 1)] = 3$ degrees of freedom for testing the fit of this model and $G_2^2 = 73.6$ is significantly different from the value which would be obtained if the model did in fact fit the data.

Even though this model does not fit the data it has improved the situation somewhat over the model of mutual independence. Let us evaluate the improvement. The difference between the values of G^2 for mutual independence and for this model is 6.0, and the difference in degrees of freedom between

Table 5.4.7. Goodness of Fit of the Model of the Independence of Death Type from Age and Medical Specialty Jointly

MD	Death	Age	Observed	Expected	$O \ln(O)$	$O \ln(m)$
1	1	1	205	169.0	1091.2	1051.6
1	1	2	148	202.4	739.6	785.9
1	2	1	352	387.6	2064.0	2097.9
1	2	2	520	464.1	3252.0	3192.9
2	1	1	42	19.1	157.0	123.9
2	1	2	9	11.8	19.8	22.2
2	2	1	21	43.8	64.0	79.4
2	2	2	30	27.1	102.0	99.0
Total					7489.6	7452.8

$G_2^2 = 2[7489.6 - 7452.8] = 73.6$, $df_2 = [(I - 1)(JK - 1)] = 3$.

Table 5.4.8. Standardized Deviates for the Model of the Independence of Death Type from Age and Medical Specialty Jointly

		Age	
		25–54	55–79
Non-psychiatry	Suicide	2.7	−3.9
	Other	−1.8	2.6
Psychiatry	Suicide	5.2	−0.8
	Other	−3.4	0.6

these two models is 1:

$$G_1^2 - G_2^2 = 79.6 - 73.6 = 6.0$$

$$df_1 - df_2 = 4 - 3 \qquad = 1.$$

Since the critical value of G^2 with 1 degree of freedom is 3.84 we conclude that the observed value of 6.0 is a significant improvement. This means that the inclusion of an interaction term expressing the dependence between age and medical specialty enhances the fit between the model and the observations in a significant manner. But clearly the model of the independence of death type from the joint distribution of age and medical specialty, even though it is better than the model of mutual independence, is still a long way from being an adequate model of the data.

The standardized deviates of expected values from observations are in Table 5.4.8. Note the large excess of suicides among young psychiatrists.

You should note very carefully that this evaluation of the improvement of fit between different models is a luxury which is provided only by the use of log-linear models as opposed to traditional chi-square tests of goodness of fit. This is a very great advantage in the process of examining data as well as testing complex models. For specific tests of specific hypotheses the values of G^2 and chi-square are very close together and there is no prior logic to establish a preference for one or the other when N is "about four or five times the number of cells" (Fienberg, 1980, 173), though the flexibility and intuitive simplicity of linear models is certainly noteworthy. The ability to fit a model in a stepwise fashion to a set of observations is a very powerful technique for data exploration.

5.4.1.1.3. Conditional Independence of One Variable from Another Given the Third

Next we evaluate the model of conditional independence of age from death type given medical specialty. The model is expressed by

$$\ln(\mu_{ijk}) = u + u_{1(i)} + u_{2(j)} + u_{3(k)} + u_{12(ij)} + u_{13(ik)}. \qquad (5.20)$$

The closed expression for the estimates of the expected values is

$$m_{ijk} = (x_{ij+})(x_{i+k})/(x_{i++}) \qquad (5.21)$$

Table 5.4.9. Marginals Required by the Model of the Conditional Independence of Age from Death Type Given Medical Specialty

Age by medical specialty marginals (x_{i+k})		
	Age	
	25–54	55–79
Non-psychiatry	557	668
Psychiatry	63	39

Death type by medical specialty marginals (x_{ij+})		
	Death type	
	Suicide	Other
Non-psychiatry	353	872
Psychiatry	51	51

Medical specialty marginals (x_{i++})	
Non-psychiatry	1225
Psychiatry	102

which on transformation becomes

$$\ln(m_{ijk}) = \ln(x_{ij+}) + \ln(x_{i+k}) - \ln(x_{i++}) \tag{5.22}$$

where it is assumed that

$$u_{23(jk)} = u_{123(ijk)} = 0.$$

Note that now two sets of two-way marginals and one set of one-way marginals (Table 5.4.9) are required: the marginal of age by medical specialty and the other of death type by medical specialty.

In Table 5.4.10 the expected values under the model of the conditional independence of age from death type given medical specialty are obtained. And in Table 5.4.11 these expected values are evaluated with regard to goodness of fit to the observed values. Note that there are $K(I-1)(J-1) = 2$ degrees of freedom for the test and that $G_3^2 = 56.8$. This outcome is very

Table 5.4.10. ln of Expected Values under the Model of the Conditional Independence of Age from Death Type Given Medical Specialty

MD	Death	Age	$u_{12(ij)}$ $\ln(x_{ij+})$	$u_{13(ik)}$ $\ln(x_{i+j})$	$u_{1(i)}$ $\ln(x_{i++})$	$\ln(m_{ijk})$
1	1	1	5.87	6.32	7.11	5.08
1	1	2	5.87	6.50	7.11	5.26
1	2	1	6.77	6.32	7.11	5.98
1	2	2	6.77	6.50	7.11	6.16
2	1	1	3.93	4.14	4.62	3.45
2	1	2	3.93	3.66	4.62	2.97
2	2	1	3.93	4.14	4.62	3.45
2	2	2	3.93	3.66	4.62	2.97

Table 5.4.11. Goodness of Fit of the Model of Conditional Independence of Age from Death Type Given Medical Specialty

MD	Death	Age	Observed	Expected	$O \ln(O)$	$O \ln(x^*)$
1	1	1	205	160.8	1091.2	1041.4
1	1	2	148	192.5	739.6	778.5
1	2	1	352	395.4	2064.0	2104.9
1	2	2	520	473.4	3252.0	3203.2
2	1	1	42	31.5	157.0	144.9
2	1	2	9	19.5	19.8	26.7
2	2	1	21	31.5	64.0	72.5
2	2	2	30	19.5	102.0	89.1
Total					7489.6	7461.2

$G_3^2 = 2[7489.6 - 7461.2] = 56.8$, $\mathrm{df}_3 = K(I-1)(J-1) = 2$.

Table 5.4.12. Standardized Deviates for
the Model of Conditional Independence
of Death Type from Age Given Medical
Specialty

		Age	
		25–54	55–79
Non-psychiatry	Suicide	3.5	−3.2
	Other	−2.2	2.0
Psychiatry	Suicide	1.9	−2.4
	Other	−1.9	2.4

improbable if the model is true. You will note, however, that the value of G_3^2 for the fit of this model is smaller than the value of G_2^2 for the model of the independence of death type from age and medical specialty jointly. So it would appear that some further improvement in model fit has been obtained by including a term for the death type by medical specialty interaction. The improvement is measured as

$$G_2^2 - G_3^2 = 73.6 - 56.8 = 16.8$$

$$df_2 - df^3 = 3 - 2 \qquad = 1.$$

The difference between the two calculated values of G^2 is 16.8 with one degree of freedom. This indicates that a significant improvement has been obtained. And in Table 5.4.12 the standardized deviates for the model of conditional independence of death type from age given medical specialty are presented.

5.4.1.1.4. Pairwise Relations among the Three Variables

If we now add the term for the interaction of death type with age then you will note that all possible two-way interactions are included in the model simultaneously. The model is expressed as

$$\ln(\mu_{ijk}) = u + u_{1(i)} + u_{2(j)} + u_{3(k)}$$
$$+ u_{12(ij)} + u_{13(ik)} + u_{23(jk)}.$$

(5.23)

Note that it is assumed that $u_{123(ijk)} = 0$.

First refer back to equations 5.18 and 5.21. All these equations are referred to as closed equations for the expected values. Unfortunately no similar equation exists for the expected values for this particular model. So an entirely different strategy must be adopted for determining the expected values. Several techniques are available, but the one that we will use here is that developed by Fienberg (1977, 1980) called Proportional Iterative Fitting. Basically what this means is that one starts with a guess as to the value in question and successively, that is iteratively, refines this guess. The particular refinement technique that is used is the ratio of the observed value to the "current" estimate.

Any technique involving iteration almost by definition requires a computer. The reason is that there is simply too much arithmetic to be done for modern students. It is a brute force kind of device for obtaining estimates for which there is no specific closed equation. All computer programs which will perform these tests of various models on contingency tables use some kind of iterative technique in order to obtain the expected values. Only the proportional iterative fitting technique will be illustrated here. An iterative technique is said to converge to a specific value when the difference between successive estimates of the value is smaller than a criterion.

Commonly the initial values are set to 1. The expected values, then, on the first step of the process are all 1. The next step is to obtain estimated expected values in a step-by-step manner until the estimates converge. It is certainly possible that the estimates will not converge, in which case there is no solution. The mechanics of the process are laid out as

$$\text{Initial values: } x_{ijk}^{(0)} = 1.$$

Then for $n = 0, 1, 2, \ldots$

$$\text{Step 1:} \quad x_{ijk}^{(3n+1)} = \frac{x_{ij+}}{x_{ij+}^{(3n)}} x_{ijk}^{(3n)}$$

$$\text{Step 2:} \quad x_{ijk}^{(3n+2)} = \frac{x_{i+k}}{x_{i+k}^{(3n+1)}} x_{ijk}^{(3n+1)}$$

$$\text{Step 3:} \quad x_{ijk}^{(3(n+1))} = \frac{x_{+jk}}{x_{+jk}^{(3n+2)}} x_{ijk}^{(3n+2)}. \tag{5.24}$$

Notice that each of the steps adjusts all of the estimates for the ratio of an observed set of marginals to a current estimate of their value. Since there are three sets of two-way marginals in this particular model, there are three steps in the adjustment procedure. Then these same three steps are performed repeatedly until convergence is achieved. Ordinarily convergence is quite rapid.

The necessary marginals are presented in Table 5.4.13.

In Table 5.4.14 the estimates of the expected values are displayed for each step in the process. Note that convergence is effectively achieved by the third cycle of iteration. The test for goodness of fit of this model is presented in Table 5.4.15. You will note that the value of G^2 has been dramatically reduced, $G_4^2 = 6.4$ with $(I-1)(J-1)(K-1) = 1$ degree of freedom. This computed value of G^2 is still significant but clearly the addition of the term for the interaction of age by death type has improved the fit greatly. Since the calculated value of G^2 is significantly large, this model does not fit the data.

Table 5.4.13. Marginals Required by the Model of All Pairwise Relations

Two-way marginals		

Specialty by death type

	Death	
	Suicide	Other
Non-psychiatry	353	872
Psychiatry	51	51

Specialty by age

	Age	
	25–54	55–79
Non-psychiatry	557	668
Psychiatry	63	39

Death type by age

	Age	
	25–54	55–79
Suicide	247	157
Other	373	550

Table 5.4.14. Convergence of Estimates of the Expected Values under the Model AD, DF, AF

			Iteration			
MD	Death	Age	1	2	3	6
1	1	1	211.6	210.5	210.7	210.7
1	1	2	139.5	142.4	142.3	142.3
1	2	1	345.4	346.5	346.3	346.3
1	2	2	528.5	525.6	525.7	525.7
2	1	1	37.9	36.4	36.3	36.3
2	1	2	15.4	14.8	14.7	14.7
2	2	1	25.1	26.6	26.7	26.7
2	2	2	23.6	24.3	24.3	24.3

Table 5.4.15. Test of Goodness of Fit of the Model with All
Pairwise Marginals

MD	Death	Age	Observed	Expected	O ln(O)	O ln(m)
1	1	1	205	210.7	1091.2	1096.8
1	1	2	148	142.3	739.6	733.8
1	2	1	352	346.3	2064.0	2058.3
1	2	2	520	525.7	3252.0	3257.7
2	1	1	42	36.3	157.0	150.9
2	1	2	9	14.7	19.8	24.2
2	2	1	21	26.7	64.0	69.0
2	2	2	30	24.3	102.0	95.7
Total					7489.6	7486.4

$G_4^2 = 2[7489.6 - 7486.4] = 6.4$, $df_4 = (I - 1)(J - 1)(K - 1) = 1$.

Table 5.4.16. Standardized Deviates for
the Model of All Pairwise Marginals

		Age	
		25–54	55–79
Non-psychiatry	Suicide	−0.4	0.5
	Other	0.3	−0.2
Psychiatry	Suicide	0.9	−1.5
	Other	−1.1	1.2

This means that the only adequate fit is the model which includes the three-way interaction term. This is referred to as the saturated model. It is saturated because it includes all terms in equation 5.16.

The standardized deviates for the fit of all pairwise marginals is presented in Table 5.4.16. The largest deviation, in absolute value, in the table is observed among psychiatrists.

In Table 5.4.17 is a summary of the results to this stage. (There are some minor differences between the log-likelihood ratio chi-squares which are reported and those which we have obtained in the text. The results in Table 5.4.17 were obtained by computer and should, therefore, be considered more accurate than those we have done with a calculator and which include a large amount of rounding error.) Model #1 in Table 5.4.17 is the model of mutual independence. Model #2 is the model of the independence of death type from age and medical specialty jointly; and model #3 is the model for the independence of age and death type conditional on field of specialty. Model #4 includes all pairwise marginals simultaneously. In the right-most two columns

Table 5.4.17. Summary of Results

Model	G^2	df	G^2 Difference	df Difference
1. A, F, D	79.6	4		
2. AF, D	69.3	3	10.3	1
3. AF, DF	50.8	2	18.5	1
4. AF, DF, AD	6.5	1	44.3	1

A age, F medical specialty, D death type.

of this table is presented the difference in observed values of the log-likelihood ratio chi-square for successive models and the degrees of freedom for the difference. Notice that the inclusion of the term for the interaction of age by medical specialty results in a decrease of G^2 of 10.3. This is the measure of the amount of improvement in goodness of fit which is achieved by including this interaction term relative to that which assumed no interactions. Continuing, then, we note that the inclusion of a term for the interaction of death type by field of medical specialty improves the fit by 18.5. Finally, the inclusion of the term for age by death type improves the fit over model #3 by 44.3. All the values of G^2 in Table 5.4.17 are significant at our chosen 0.05, 5%, level of significance; so none of the models fit the data adequately, and each increment in model complexity improves the fit significantly.

In the context of the research program with which we began, the terms for the interaction of age by medical specialty and the terms for death type by medical specialty should be considered as nuisance variables. That is to say, the original concern focussed on the interaction of age by death type. The research question of interest at the moment, then, is whether #4 in Table 5.4.17 is a better test of the research hypothesis than is the test that was performed on Table 5.3.7. Recall that the value of G^2 there was 48.7 with 1 degree of freedom, indicating that age and death type are definitely not independent of each other in these data. But those data contained a great deal of noise. We suspect that medical specialty should have some effect on death type choice. When the data are structured as in Table 5.3.6 some of this noise is removable. For example, consider model #3 in Table 5.4.17. That is the model which includes the sampling artifact of age by medical specialty and the death by medical specialty interactions. A more powerful test of the significance of the interaction of age by death type is provided by the removal of these nuisance factors. Model #4 accomplishes this, particularly when it is compared to model #3.

Recall that the research hypothesis for this work asserted that suicide frequency would increase with age. We have reached a point of considerable confidence that there is some relationship between suicide frequency and age (within the population of all U.S. physicians) but we still need to make a

Table 5.4.18. Estimated
Probabilities of Suicide by Age
for Medical Specialty

	Age	
	25–54	55–79
Overall	0.40	0.22
Non-psychiatry	0.37	0.22
Psychiatry	0.67	0.23

statement about the direction of the effect. We have successfully rejected the null hypothesis that suicide frequency is independent of age and now we ask whether it increases or decreases with age. We shall approach this problem intuitively rather than rigorously. Consider Table 5.4.18. There are presented the estimated overall probabilities of suicide in line 1 taken from Table 5.3.5 and in lines 2 and 3 are the estimated probabilities for the medical specialties taken from Table 5.3.6. Recall that it is not possible for us to be concerned with specific rates of suicide. Rather we are concerned with, for example, differences between age categories. In line 1 you will note that the change in suicide frequency with aging is devisively not positive. Lines 2 and 3 provide more sensitive comparisons between age groups, since the factor of medical specialty has been removed. Also recall that the data for line 1 includes all noise factors. You should be struck by the fact that in no case in Table 5.4.18 does the suicide frequency, expressed as the proportion of suicides, increase with age. So we are able to reject not only that part of the null hypothesis which specifies no age effect, but also to reject the other part which asserts that the frequency should increase with age. This constitutes sufficient evidence to allow rejection of the theory from which the hypothesis was obtained.

5.4.1.2. BLOOD GROUPS, ILLNESS AND CITY

In Section 5.3.1.1 the relationship of some blood group types to illness was considered. The data are also structured by city (Woolf, 1955; Dixon et al., 1981), as in Table 5.4.19.

The odds for type O are displayed in Figure 5.8 by condition and city. There you may observe that the odds are always greater for blood group type O than for type A; but the cities do not seem to be homogeneous. The results of fitting a set of possible models are presented in Table 5.4.20.

As usual, the model of mutual independence is tested first. Clearly it does not fit the data. Note that the addition of a term expressing the interaction of illness and city, IC, reduces the value of G^2 by 670.88 with 2 degrees of

Table 5.4.19. Blood Group, by Illness, by City

City	Blood Group	Illness Ulcer	Control	Total
London	O	911	4578	5489
	A	579	4219	4798
	Total	1490	8797	10287
Manchester	O	361	4532	4893
	A	246	3775	4021
	Total	607	8307	8914
Newcastle	O	396	6598	6994
	A	219	5261	5480
	Total	615	11859	12474
Total				31675

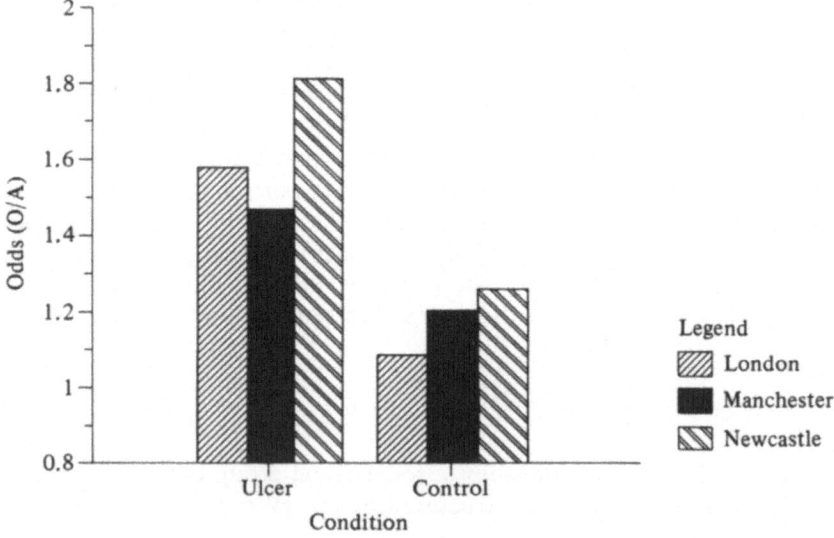

Figure 5.8

freedom. This is apparently the strongest effect, i.e. makes the biggest difference, among the single pairwise interaction models. So the process arrives at model 3. From there, the next largest effect seems to be the addition of a term for the interaction of blood group and illness, IB (model 5). Including this term produces a reduction in G^2 of 53.45 with 1 degree of freedom. Finally

Table 5.4.20. Summary of
Models Fitted to the Blood
Group, Illness, and City Data

Model	df	G^2
1. I, B, C	7	754.45
2. I, BC	5	737.76
3. B, IC	5	83.57
4. C, IB	6	700.99
5. IB, IC	4	30.12
6. IC, BC	3	66.90
7. BC, IB	4	684.28
8. IB, IC, BC	2	2.97

B: blood group, I: illness, C: city.

Table 5.4.21. Expected Values Produced by the
Model Using All Pairwise Interactions for the
Blood Group, Illness, and City Data

City	Blood Group	Illness	
		Ulcer	Control
London	O	898.3	4590.7
	A	591.7	4206.3
Manchester	O	378.5	4514.5
	A	228.5	3792.5
Newcastle	O	391.2	6602.8
	A	223.8	5256.2

this inclusion of the last of the pairwise terms reduces G^2 by 27.15 with 2 degrees of freedom and produces an adequate model of the data (model 8). Note that the model which includes only pairwise interations is much simpler than that which requires the simultaneous interaction of all three variables and, so, is a great advance. Also, however, note that each term would probably be of interest separately to different kinds of investigators. For example, a study of genetic drift would likely focus on the BC term (Why are there differences among cities for the blood group type?), an epidemiologist might attend the IC term (Why is the frequency of illness different among cities?), and an evolutionist would be attracted by the IB term (Why is the frequency of illness different for the blood group types?).

The expected values produced by model 8 are presented in Table 5.4.21.

5.4.2. The 2 × 2 × 2 × 2 Table: Sickle-Cell Trait, Age, Sex, and City

In this section we consider an extension of the previous results to a 4-way table. (In principle, higher dimensions can be analyzed, but as a practical matter a 5-way table is about the useful limit.) In order to motivate this, we shall consider some data on sickle-cell trait in the U.S.

One of the things that has received the attention of evolutionary theorists for some time is what happens to a particular allele when the selective pressure maintaining it is removed. An example is the sickle-cell trait. This trait, as you know, is highly deleterious in the homozygous state, and so under ordinary circumstances one would expect it to be eliminated rather quickly. However, in West Africa and other places, the frequency of the trait is maintained in rather high frequencies through its role in protecting against malaria. When African populations are removed to the New World they enter an area where malaria, while present in the southeastern United States, is certainly not a major health hazard. Effectively then one may consider this to be a population for which the selective pressure for high frequencies of sickle-cell trait has been removed. Until very recently, recognizing that the trait is deleterious in a double dose, most evolutionary theorists were of the opinion that a population snapshot at any given point in time would exhibit a declining frequency with age. The expectation clearly has been that natural mortality will tend selectively to eliminate trait carriers from the population.

In the 1970s a large number of sickle-cell screening clinics were operated throughout the United States. The purpose was to locate individuals who had a copy of the trait in order to advise them of this fact. There was also concern by public health officials that these individuals might be at somewhat greater

Table 5.4.22. Genotype Frequency by Age, Sex, and City

Sex	Genotype	City	Age 01–20	Age GT 20
Male	Normal	Tampa	4026	821
		New York	7003	1210
	Sickler	Tampa	427	95
		New York	522	108
Female	Normal	Tampa	5083	2235
		New York	8050	2204
	Sickler	Tampa	507	268
		New York	561	180

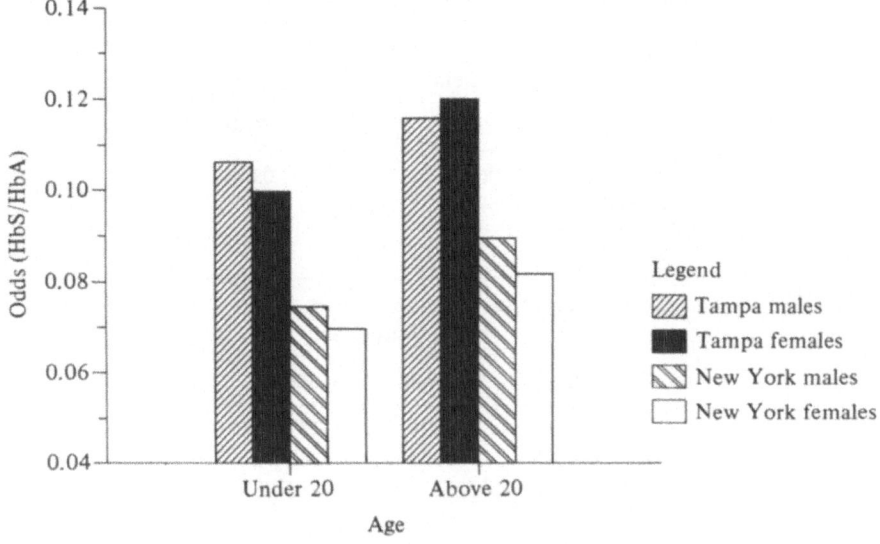

Figure 5.9

Table 5.4.23. Genotype by Age
Marginals for Sickle-Cell Frequency

	Genotype		
Age	Normal	Sickler	P(S)
01–20	24162	2017	0.077
GT 20	6470	651	0.101

$G^2 = 15.2$, df = 1.

risk for a variety of potential problems. Two large screening clinics were operated in Tampa, Florida, and New York City.

In Table 5.4.22 are the frequencies of carriers of the trait and normals displayed by age, sex, and location. The odds for sickle-cell by age are presented in Figure 5.9. The evolutionary hypothesis of interest is that the trait frequency should be lower in the older age group than it is in the younger age group. In Table 5.4.23 are the frequencies of the genotypes by age. The value of G^2 for this table is 15.2 with one degree of freedom which is a significant value. Also note that the probability of the sickler genotype among the young is 0.077, while among the older age group it is 0.101. So age and genotype are clearly not independent, but the direction of the difference in frequencies between the two age groups is not that which was predicted.

Table 5.4.24. Genotypes by Sex

a. Genotype by sex marginals

Sex	Genotype		P(S)
	Normal	Sickler	
Male	13060	1152	0.081
Female	17572	1516	0.079

$G^2 = 0.29$, df $= 1$.

b. Age by genotype given sex marginals

	Age	Genotype		P(S)
		Normal	Sickler	
Male	01–20	11029	949	0.079
	GT 20	2031	203	0.091

$G^2 = 3.3$, df $= 1$.

| Female | 01–20 | 13133 | 1068 | 0.075 |
| | GT 20 | 4439 | 448 | 0.092 |

$G^2 = 13.1$, df $= 1$.

In Table 5.4.24a are the genotype by sex marginals. Note that the value of G^2 for this table is non-significant. This is reflected in the relative frequencies of the trait carriers for male and female which are almost identical. In part b of this table are the age by genotype marginals for each sex. Note that for males the $G^2 = 3.3$, is smaller than the critical value of 3.84, with one degree of freedom. Consequently we would assert that there is no difference in frequency by age for males. In the second part are the results for females. $G^2 = 13.1$ is significant so we are confident of significant age differences among females. Note that the direction of the difference is opposite that predicted.

In Table 5.4.25a are the genotype by city marginals. There is a clear difference between Tampa and New York in the overall frequency of the trait. In part b of that table is displayed the genotype by age marginals for each city. Note that while the age effect is significant in both locations, most of the difference between cities is in the overall relative frequency, and very little in the age effect. Still, however, the direction of the difference is opposite that predicted.

Refer back to Table 5.4.24. There we observed that overall there is no difference between male and female with regard to the relative frequency of the trait. However, when this result was elaborated in part b of that table by discriminating between age categories it was observed that there is no significant difference between the age categories for males but a highly significant

Table 5.4.25. Genotypes by City

a. Genotype by city marginals

| | Genotype | | |
City	Normal	Sickler	P(S)
Tampa	12165	1297	0.096
New York	18467	1371	0.069

$G^2 = 79.5$, df $= 1$.

b. Genotype by age given city marginals

| | Age | Genotype | | |
		Normal	Sickler	P(S)
Tampa	01–20	9109	934	0.093
	GT 20	3056	363	0.106

$G^2 = 5.0$, df $= 1$.

| New York | 01–20 | 15053 | 1083 | 0.067 |
| | GT 20 | 3414 | 288 | 0.078 |

$G^2 = 5.2$, df $= 1$.

difference between age categories for females. This is primarily the result of there being a larger deficit of trait carriers among young females than is true for males, associated with there being correspondingly greater excess of older trait carriers among females than among males.

By now it is clear that there is a large number of sampling artifacts present in these data. Recall that any effect other than that related to the relationship between age and genotype is considered to be noise. Specifically this means that any difference in relative frequencies for the age categories by sex, any difference in the age categories for cities, any difference in the relative frequencies for the sexes by the cities or any combination of these should be removed from the data in so far as it is possible. This is accomplished by "fixing the marginals." When marginals are fixed this is equivalent to declaring a variable to be an independent, or explanatory, variable. In the case at hand the variable is a complex one, that is, it actually involves three observational variables (age, sex, city). So we must remove the effect of the interactions between these three observational variables. This is accomplished by fixing that set of three-way marginals. Then the variablility which remains, residual variability, in the relative frequencies of trait carriers for the age categories is that which is of interest. Notice that this process of the removal of variability is not restricted to the three-way marginals for age, sex, and city. We should remove all sources of variability which in some sense contaminate the relationship of concern. In Table 5.4.26 are presented some tests of significance on

Table 5.4.26. Goodness of Fit Tests for Some Models for the
Effect of Age, Sex, and City on Genotype

Model	G^2	P	df	G^2 Difference	df Difference
1. ASC, G	92.2	0.0	7		
2. ASC, GC	12.6	0.04	6	79.6	1
3. ASC, GC, GA	2.5	0.78	5	10.1	1

A: age, S: sex, C: city, G: genotype.

residual variability after the three-way interaction for age, sex, and city has
been removed. Model #2 there tests for the significance of the difference
between relative frequencies for the genotypes between the cities after some
of the noise has been removed. Notice that $G^2 = 12.6$ for model #2 is
significant and so we are confident that this model is not adequate. We may
therefore conclude that the interaction between genotype frequencies and
cities is also noise. In model #3 the two-way interaction for genotype by age
has been added to model #2. Note that $G^2 = 2.5$ is insignificant with 5
degrees of freedom. This indicates that model #3 fits the data adequately. We
confidently conclude then that there is an age by genotype interaction in the
residual variability after the noise has been removed. Observing the results
that were obtained earlier we are also confident that the direction of this effect
is the opposite of that which is predicted by the theory with which we began.
And in fact an examination of the parameters of the model confirms this
observation.

It may have occurred to you that the proper models for comparison are:
(1) ASC, GC, GS, and (2) ASC, GC, GS, GA. In this case the difference in complexity which results by including the interaction term with sex is relatively
minor when compared to the accepted model. As a general rule we seek the
simplest possible model which provides an adequate test of the hypothesis.
The model which has been accepted is simpler than that which includes an interaction term GS and is, therefore, preferred. The process of selecting a model
to fit the data is at least partially artful. There are no firm rules which can be
presented which assure a proper decision. Guidelines such as "simplicity" and
"adequacy" are necessarily invoked.

Interpreting this result—a positive relationship between age and genotype
frequency—is straightforward. The only evolutionary force which could produce this observation is differential fertility. Parents with normal hemoglobin
produce more offspring than do those with the sickle-cell trait. This fertility
differential is greater than the mortality differential so the slope of the response
frequency curve is positive.

We shall drop the matter here, and leave the intriguing questions related
to sex differences for another time. It appears to be the case that the difference
between the rate of production of male trait carriers is not greater than male
mortality.

5.5. Special Topics

5.5.1. Test of a Markovian Hypothesis

The initial question needing attention is whether the transition probabilities are changing through time. All the machinery of Section 4.3.2 assumes that the matrix of transition probabilities contains only constants. In order that this determination be made, the matrix of probabilities must be estimated at (at least) two points in time. This may be considered as a problem in 3-dimensions, the first two describe the matrix, and so are equal, and the third is time. The log-linear model corresponding to the question of constant probabilities is

$$u + u_{1(j)} + u_{2(j)} + u_{3(k)} + u_{12(ij)} + u_{13(ik)}$$

(Bishop, Fienberg, and Holland, 1977, 265). The term $u_{13(ik)}$ must not be zero because the rows of each matrix in the series must sum to 1.0. No such constraint exists for the columns of the matrices, so $u_{23(jk)} = 0$. Of course, $u_{123(ijk)} = 0$. If this model fits, the observations are not incompatible with the hypothesis of a stationary Markov chain.

5.5.2. Test of a Causal Hypothesis

These concepts will be illustrated with reference to the blood group, illness, and city data of Section 5.4.1.2. Recall that caustion can never be observed directly. It is possible, however, to make observations which will allow the rejection of a theoretical hypothesis which is obtained by assuming a particular causal relationship among the variables. The basic tool for describing causal assumptions is the path diagram. By definition, the causal relationship is asymmetrical. If A causes B, then it cannot be the case that B causes A simultaneously. (Note that the special, though very common, case of feedback is characterized by a *sequence* of reversals of the direction of the effect.) It is only necessary that the connectedness be specified between any two variables in order that a complete set of path diagrams be constructed. The set of connections among three variables is in Figure 5.10. When a diagram of connectedness is augmented with directional indicators, it is called a path diagram. Let us list all the path diagrams in each of the groups above. In the first the constraint is that all pairs are present. Specifically we may write

1.1	AB, BC, CA
1.2	AB, BC, AC
1.3	AB, CB, CA
1.4	AB, CB, AC
1.5	BA, BC, CA
1.6	BA, BC, AC
1.7	BA, CB, CA
1.8	BA, CB, AC

1. All pairs of variables connected

2. Two of three pairs of variables connected

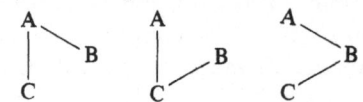

3. One of three pairs of variables connected

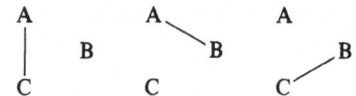

4. No connections

Figure 5.10

where the sequence indicates the direction of the effect that is, "AB" means "A causes B." In the second group the constraint is that only two pairs are present,

2.1.1	AC, AB
2.1.2	AC, BA
2.1.3	CA, AB
2.1.4	CA, BA
2.2.1	AC, BC
2.2.2	AC, CB
2.2.3	CA, BC
2.2.4	CA, CB
2.3.1	AB, BC
2.3.2	AB, CB
2.3.3	BA, BC
2.3.4	BA, CB

In row three the constraint is that only one pair is present:

3.1.1	AC
3.1.2	CA
3.2.1	AB
3.2.2	BA
3.3.1	BC
3.3.2	CB

And in row four there are no pairs.

Note that there are 27 path diagrams for the possible relationships among three variables. The hypotheses which are testable with a log-linear model are of the form: "the correct path diagram is one of group $\{1, 2, 3, 4\}$." All models within a group are statistically equivalent. Unless the theory is quite specific about the causal relations among variables it is not particularly meaningful to determine the most likely group for the model.

Fienberg (*ibid.*, 91ff) shows that the following log-linear models are associated with each group of path diagrams:

Group	Log-linear model
1	$u_{ABC} = 0$
2.1	$u_{BC} = u_{ABC} = 0$
2.2	$u_{AB} = u_{ABC} = 0$
2.3	$u_{AC} = u_{ABC} = 0$
3.1	$u_{AB} = u_{BC} = u_{ABC} = 0$
3.2	$u_{AC} = u_{BC} = u_{ABC} = 0$
3.3	$u_{AC} = u_{AB} = u_{ABC} = 0$
4	$u_{AB} = u_{BC} = u_{CA} = u_{ABC}$ $= 0$

For example, consider the blood group, disease, and city model of Section 5.4.1.2. Recall that a model of group 1 was accepted there, i.e.

$$u_{IBC} = 0$$

where I: illness, B: blood group, C: city. There is no ambiguity about causal direction for these data. The term u_{IB} is interpreted as "[blood group] causes [condition]." The reverse, fortunately, is nonsense. Also the term u_{IC} is interpreted as "[city] causes [condition];" and the term u_{BC} is "[city] causes [blood group]." To reverse any of these pairwise relations creates gibberish. Recognizing that, however, does not solve the problems. In fact, none of the terms is meaningful without prior theoretical justification. For example, the term "[blood group] causes [conditions]" seems to require a medical explanation, the term "[city] causes [condition]" needs an epidemiological explanation, and "[city] causes [blood group]" needs an explanation from evolutionary biology.

It is essential to realize that the matter of causation is not amenable to a statistical resolution. A theory which asserts a causal connection may only be supported or rejected.

CHAPTER 6

Summary and Conclusions

This entire effort should be treated as an analog of the sketch of a building on the back of an envelope. The intended construction is anthropology. There is only the suggestion of materials—that comes much later. Even the overall outline of the edifice is little more than a few shaky scratches—hinting at the thing to come. Many more versions will be proposed before delivering the concept to the draftsmen. Anthropologists have spent a great deal of time discussing the bricks with no clear idea of what is on order. This is the way termites work. Eventually termites do produce vast and wonderful mounds, but the human brain can improve on this procedure. If not, we should all wear masks when accepting a paycheck. A comprehensive theory with deductive power is required to give focus and direction to technical effort. No termite has ever designed a mound. It is still uncertain whether any human has ever designed a theory of anthropology. So the discipline is being built willy-nilly with neither long-term objectives nor day-do-day directives.

An exercise that has produced great consternation among many of my students is to find some examples in the literature of each of the four argument structures presented in Chapter 2. Attending structure only and ignoring, largely, the general failure to deduce hypotheses, logically valid arguments are distressingly rare. It is quixotic to search for examples of valid arguments which satisfy both conditions of a good test. There are occasional claims, with associated empirical justification, that, a sequence of observations is adequately described by a model of some type or another. In order that these efforts rise above the descriptive level two additional steps are necessary: (1) it must be shown that if a model of this type is appropriate the observations would result, and (2) no other type of model could produce them. These steps require an understanding of dynamics at a level foreign to most anthropologists. They are, nonetheless, essential to the development of a mature science.

Statistical tests of true hypotheses—not guesses, or hunches, or such ilk—are an essential ingredient of the enterprise. This is especially true for a nascent science such as anthropology. At best, at this stage of development of our discipline, all models can be only approximate. Statistical tests are the only way to decide if a specific, approximate, model has potential explanatory power or not. Usually, the decision is negative. One of the chronic failures of anthropology becomes evident when the decision is tentatively positive. There is no sequel, no follow-up. The model is not investigated to exhaust all its potential. It is not refined and made more precise. Partial success kills anthropological theory. This is an incomprehensible state of affairs. It is tempting to blame professional, career, concerns and entrenched professional arbiters. A small success is a fragile thing and few of us are ready to go to the wall with only that for defense. And few of us possess the "insanity" of Mendel or Darwin or Hamilton. Many disciplines have been resurrected by marginal individuals from outside the ranks of the professional elite. There is no reason to think that anthropology is any different from, say, pre-Darwinian biology.

Historically, anthropology in North America has had a strong preference for documenting the bewildering variety of solutions to a few human "needs" in the interest of showing that one solution is as good as another. This effort is as variable as the solutions. Many have observed that the writing of ethnography is the anthropologist's TAT test. This idiosyncratic fascination with data has produced a body of scientifically useless, though engrossing, literature. A change in orientation is overdue.

When it occurs, it will have the components outlined here. Valid argument structure will be fundamental without the baroque lapses into sophistry. Most attention will be directed to the model of the theory. Through rigorous argument, then, a hypothesis will be shown to be a deductive consequence of the theory. Finally, the hypothesis will be tested statistically. My confidence stems from the simple recognition that these are the components of a mature science.

Matrix Manipulation

A matrix is a rectangular array of numbers. The dimensions of a matrix are r × c where r is the number of rows and c the number of columns. Matrices will be identified by uppercase boldface, e.g.

$$\mathbf{A} = \begin{bmatrix} a_{11} & a_{12} & \cdots & a_{1c} \\ a_{21} & a_{22} & \cdots & a_{2c} \\ \vdots & & & \\ a_{r1} & & & a_{rc} \end{bmatrix}.$$

Two matrices are equal when all their elements are equal.

Matrix addition is defined only for matrices of the same dimensions, e.g.

$$\mathbf{Q} = \mathbf{A} + \mathbf{B}$$

$$= \begin{bmatrix} a_{11} & a_{12} & a_{13} & a_{14} \\ a_{21} & a_{22} & a_{23} & a_{14} \\ a_{31} & a_{32} & a_{33} & a_{34} \end{bmatrix} + \begin{bmatrix} b_{11} & b_{12} & b_{13} & b_{14} \\ b_{21} & b_{22} & b_{23} & b_{24} \\ b_{31} & b_{32} & b_{33} & b_{34} \end{bmatrix}$$

$$= \begin{bmatrix} a_{11}+b_{11} & a_{12}+b_{12} & a_{13}+b_{13} & a_{14}+b_{14} \\ a_{21}+b_{21} & a_{22}+b_{22} & a_{23}+b_{23} & a_{24}+b_{24} \\ a_{31}+b_{31} & a_{32}+b_{32} & a_{33}+b_{33} & a_{34}+b_{34} \end{bmatrix}.$$

When one of the dimensions of a matrix is 1 it is called a vector; e.g.

$$\mathbf{a} = \begin{bmatrix} a_{11} \\ a_{21} \\ \vdots \\ a_{r1} \end{bmatrix}, \qquad \mathbf{b} = (b_{11} \quad b_{12} \quad \cdots \quad b_{1c})$$

a is (r × 1), an r element column vector; and **b** is (1 × c), a c element row vector. Vectors will be identified by lowercase boldface.

Any matrix may be multiplied by a scalar, a (1 × 1) matrix, e.g.

$$c \cdot A = \begin{bmatrix} ca_{11} & ca_{12} & \cdots & ca_{1c} \\ ca_{21} & ca_{22} & \cdots & ca_{2c} \\ \vdots & & & \\ ca_{r1} & ca_{r2} & & ca_{rc} \end{bmatrix}.$$

Two matrices may be multiplied when the "internal dimension" is the same, e.g. if **A** is (r × c) and **B** is (c × d) the product **A · B** is defined but the product **B · A** is not unless r = d. The product matrix has dimensions (r × d). Order of multiplication is relevant, which distinguishes matrix multiplication from the multiplication of numbers. Matrix multiplication is defined by the following

$$A \cdot B = Q$$

$$Q = \begin{bmatrix} q_{11} = \sum_{i,j} a_{1j}b_{i1} & q_{12} = \sum_{i,j} a_{1j}b_{i2} & q_{13} = \sum_{i,j} a_{1j}b_{i3} \\ q_{21} = \sum_{i,j} a_{2j}b_{i1} & q_{22} = \sum_{i,j} a_{2j}b_{i2} & q_{23} = \sum_{i,j} a_{2j}b_{i3} \\ q_{31} = \sum_{i,j} a_{3j}b_{i1} & q_{32} = \sum_{i,j} a_{3j}b_{i2} & q_{33} = \sum_{i,j} a_{3j}b_{i3} \end{bmatrix}.$$

For example

$$A = \begin{bmatrix} 2 & 1 & 3 \\ 0 & 1 & 2 \end{bmatrix}, \qquad B = \begin{bmatrix} a & d \\ b & e \\ c & f \end{bmatrix}$$

then

$$Q = A \cdot B = \begin{bmatrix} 2a + b + 3c & 2d + e + 3f \\ b + 2c & e + 2f \end{bmatrix}.$$

Notice that the product **B · A** is also defined

$$S = B \cdot A = \begin{bmatrix} 2a & a + d & 3a + 2d \\ 2b & b + e & 3b + 2e \\ 2c & c + f & 3c + 2f \end{bmatrix}.$$

Note that $A \cdot B \neq B \cdot A$.

The following is a modification of an example given by Emlen (1984). Suppose a population with 4 age classes, say 0–15, 15–30, 30–45, 45+ for example. There are $n_1(t)$ in age class 1 at time t, $n_2(t)$ in class 2, $n_3(t)$ in class 3 and $n_4(t)$ in class 4. Then at time t the population is described by the vector

$$\begin{bmatrix} n_1(t) \\ n_2(t) \\ n_3(t) \\ n_4(t) \end{bmatrix}.$$

Suppose the probability of surviving class 1 and entering class 2 at time $t + 1$ is p_1, of surviving class 2 and entering class 3 at time $t + 1$ is p_2, of surviving class 3 is p_3. The probability of surviving class 4 is 0. Then

$$n_2(t + 1) = p_1 n_1(t)$$
$$n_3(t + 1) = p_2 n_2(t)$$
$$n_4(t + 1) = p_3 n_3(t).$$

The number of offspring still alive after 1 time unit born to mothers in all age classes is $(0, b_2, b_3, b_4)$. These offspring constitute the first age class so that

$$n_1(t + 1) = 0n_1(t) + b_2 n_2(t) + b_3 n_3(t) + b_4 n_4(t).$$

These results may be displayed as the matrix equation

$$\begin{bmatrix} n_1(t + 1) \\ n_2(t + 1) \\ n_3(t + 1) \\ n_4(t + 1) \end{bmatrix} = \begin{bmatrix} 0 & b_2 & b_3 & b_4 \\ p_1 & 0 & 0 & 0 \\ 0 & p_2 & 0 & 0 \\ 0 & 0 & p_3 & 0 \end{bmatrix} \begin{bmatrix} n_1(t) \\ n_2(t) \\ n_3(t) \\ n_4(t) \end{bmatrix}.$$

It is instructive to perform the indicated multiplication.

$$\begin{bmatrix} n_1(t + 1) \\ n_2(t + 1) \\ n_3(t + 1) \\ n_4(t + 1) \end{bmatrix} = \begin{bmatrix} 0n_1(t) + b_2 n_2(t) + b_3 n_3(t) + b_4 n_4(t) \\ p_1 n_1(t) \\ p_2 n_2(t) \\ p_3 n_3(t) \end{bmatrix}.$$

If the matrix of coefficients is denoted $\mathbf{A}$ and the population vector $\mathbf{n}$ then we may write

$$\mathbf{n}(t + 1) = \mathbf{A}\mathbf{n}(t)$$

which is a very useful economy of notation as the number of age categories increase.

From the foregoing it is evident that the multiplication of square matrices has many, but not all, of the properties of ordinary multiplication. For example, for all real numbers

$$a(1/a) = 1$$

but the analog of the reciprocal, called the matrix inverse, does not always exist. For a matrix $\mathbf{A}$ we seek a matrix $\mathbf{A}^{-1}$ such that

$$\mathbf{A}\mathbf{A}^{-1} = \mathbf{A}^{-1}\mathbf{A} = \mathbf{I}$$

where $\mathbf{I}$ is the identity matrix,

$$I = \begin{bmatrix} 1 & 0 & 0 & \cdots \\ 0 & 1 & 0 & \cdots \\ \vdots & & & \\ 0 & 0 & 0 & \cdots\cdots & 1 \end{bmatrix}.$$

Note that only square matrices have an inverse.

Here is presented one of several methods of finding A^{-1}, called the method of cofactors. Consider the (3×3) matrix

$$A = \begin{bmatrix} a_1 & b_1 & c_1 \\ a_2 & b_2 & c_2 \\ a_3 & b_3 & c_3 \end{bmatrix}.$$

Each element in A has a minor defined as

Element	Minor
a_1	$b_2 c_3 - b_3 c_2$
a_2	$b_1 c_3 - b_3 c_1$
a_3	$b_1 c_2 - b_2 c_1$
b_1	$a_2 c_3 - a_3 c_2$
b_2	$a_1 c_3 - a_3 c_1$
b_3	$a_1 c_2 - a_2 c_1$
c_1	$a_2 b_3 - a_3 b_2$
c_2	$a_1 b_3 - a_3 c_1$
c_3	$a_1 b_2 - a_2 b_1$

When the minor is signed it is called a cofactor. The sign is determined as follows. Label the rows from top to bottom $(1, 2, 3)$, and label the columns similarly from left to right. Find the sum of the row and column label of the element. If the sum is even, the sign of the cofactor is $+$; if the sum is odd the sign is $-$.

Element	Row	Column	Sign
a_1	1	1	$+$
a_2	2	1	$-$
a_3	3	1	$+$
b_1	1	2	$-$
b_2	2	2	$+$
b_3	3	2	$-$
c_1	1	3	$+$
c_2	2	3	$-$
c_3	3	3	$+$

The cofactor of the element is identified with a capital letter

Element	Cofactor
a_1	$A_1 = +(b_2 c_3 - b_3 c_2)$
a_2	$A_2 = -(b_1 c_3 - b_3 c_1)$
a_3	$A_3 = +(b_1 c_2 - b_2 c_1)$
b_1	$B_1 = -(a_2 c_3 - a_3 c_2)$
b_2	$B_2 = +(a_1 c_3 - a_3 c_1)$
b_3	$B_3 = -(a_1 c_2 - a_2 c_1)$
b_1	$C_1 = +(a_2 b_3 - a_3 b_2)$
b_2	$C_2 = -(a_1 b_3 - a_3 c_1)$
b_3	$C_3 = +(a_1 b_2 - a_2 b_1)$

A property of the matrix called the determinant is defined as

$$\det A = a_1 A_1 + a_2 A_2 + a_3 A_3.$$

As you will see below, if $\det A = 0$, then A has no inverse.
    It can be easily shown that

$$A^{-1} = 1/\det A \begin{bmatrix} A_1 & A_2 & A_3 \\ B_1 & B_2 & B_3 \\ C_1 & C_2 & C_3 \end{bmatrix}$$

if $\det A \neq 0$. For example let

$$A = \begin{bmatrix} 1 & -q_2 & 0 \\ -O_2 & 1 & -O_1 \\ 0 & -q_1 & 1 \end{bmatrix}.$$

The matrix of cofactors is

$$\begin{bmatrix} 1 - q_1 O_1 & q_2 & q_2 O_1 \\ O_2 & 1 & O_1 \\ q_1 O_2 & q_1 & 1 - q_2 O_2 \end{bmatrix}$$

then

$$\det A = 1 - q_1 O_1 - q_2 O_2$$

which will be zero when

$$1 = q_1 O_1 + q_2 O_2.$$

Then

$$A^{-1} = \begin{bmatrix} \dfrac{1 - q_1 O_1}{1 - q_1 O_1 - q_2 O_2} & \dfrac{q_2}{1 - q_1 O_1 - q_2 O_2} & \dfrac{q_2 O_1}{1 - q_1 O_1 - q_2 O_2} \\[3mm] \dfrac{O_2}{1 - q_1 O_1 - q_2 O_2} & \dfrac{1}{1 - q_1 O_1 - q_2 O_2} & \dfrac{O_1}{1 - q_1 O_1 - q_2 O_2} \\[3mm] \dfrac{q_1 O_2}{1 - q_1 O_1 - q_2 O_2} & \dfrac{q_1}{1 - q_1 O_1 - q_2 O_2} & \dfrac{1 - q_2 O_2}{1 - q_1 O_1 - q_2 O_2} \end{bmatrix}.$$

Let us confirm that this is the inverse. If so then $AA^{-1} = I$, so we may list the elements of the product. Those elements on the diagonal must be 1, and all others 0.

$$I_{11} = \frac{1 - q_1 O_1}{1 - q_1 O_1 - q_2 O_2} - \frac{q_2 O_2}{1 - q_1 O_1 - q_2 O_2} = 1$$

$$I_{12} = \frac{-q_2}{1 - q_1 O_1 - q_2 O_2} - \frac{q_2}{1 - q_1 O_1 - q_2 O_2} = 0$$

$$I_{13} = \frac{q_2 O_1}{1 - q_1 O_1 - q_2 O_2} - \frac{q_2 O_1}{1 - q_1 O_1 - q_2 O_2} = 0$$

$$I_{21} = \frac{-O_2(1 - q_1 O_2)}{1 - q_1 O_1 - q_2 O_2} + \frac{O_2}{1 - q_1 O_1 - q_2 O_2} - \frac{O_1(q_1 O_2)}{1 - q_1 O_1 - q_2 O_2} = 0$$

$$I_{22} = -\frac{O_2 q_2}{1 - q_1 O_1 - q_2 O_2} + \frac{1}{1 - q_1 O_1 - q_2 O_2} - \frac{q_1 O_1}{1 - q_1 O_1 - q_2 O_2} = 1$$

$$I_{23} = \frac{-O_2(q_2 O_1)}{1 - q_1 O_1 - q_2 O_2} + \frac{O_1}{1 - q_1 O_1 - q_2 O_2} + \frac{-O_1(1 - q_2 O_2)}{1 - q_1 O_1 - q_2 O_2} = 0$$

$$I_{31} = -\frac{q_1 O_2}{1 - q_1 O_1 - q_2 O_2} + \frac{q_1 O_2}{1 - q_1 O_1 - q_2 O_2} = 0$$

$$I_{32} = -\frac{-q_1}{1 - q_1 O_1 - q_2 O_2} + \frac{q_1}{1 - q_1 O_1 - q_2 O_2} = 0$$

$$I_{33} = -\frac{q_1 O_1}{1 - q_1 O_1 - q_2 O_2} + \frac{1 - q_2 O_2}{1 - q_1 O_1 - q_2 O_2} = 1.$$

This demonstrates that A^{-1} is actually the inverse of A.

Conversion of the Base of Logarithms

Recall that a logarithm is an exponent. For base a, $\log_a(x)$ is the power to which a must be raised to produce x. That is, if $\log_a(x) = y$ then $a^y = x$—when a is raised to the power y, the result is x. Two common bases are 10 and $e \sim 2.7183$. Modern hand calculators will return the $\log_{10}$ usually with a key labelled "log" and the $\log_e$ with a key "ln." They do not commonly produce $\log_2$, the conversion from one base to base 2 is often required. Fortunately it is simple.

$$\log_2(x) = \log_{10}(x)/\log_{10}(2) = \log_{10}(x)/0.3010$$

$$\log_2(x) = \log_e(x)/\log_e(2) = \log_e(x)/0.6931.$$

Bayes' Theorem

In discussions of conditional probability it has become conventional to label the conditioning event as the hypothesis. Assume for example that there are two possible mutually exclusive hypotheses for an event. Bayes' theorem is a mathematically rigorous technique for obtaining the conditional probability of the hypothesis given the data. It is a straightforward exercise. Let A denote the event, H_1 and H_2 are two hypotheses. Then consider

$$P(A|H_1) = P(A \text{ and } H_1)/P(H_1)$$

$$P(A|H_2) = P(A \text{ and } H_2)/P(H_2)$$

and we are interested in an expression of the sort $P(H_1|A)$. Notice that this would be written as

$$P(H_1|A) = P(A \text{ and } H_1)/P(A).$$

The conjunction of the events A and H_1 can be obtained from

$$P(A \text{ and } H_1) = P(H_1) \cdot P(A|H_1).$$

The unconditional probability of the event A may be obtained by

$$P(A) = P(A \text{ and } H_1) + P(A \text{ and } H_2).$$

Now the desired result is given by

$$P(H_1|A) = \frac{P(H_1) \cdot P(A|H_1)}{[P(A \text{ and } H_1) + P(A \text{ and } H_2)]}.$$

Lest you be tempted to see in this formulation a mechanical salvation for science, consider that if you conclude for example that H_1 is true—because $P(H_1|A)$ is large—then you have committed the logical fallacy called affirming the consequent. Also note that this argument depends on the existence of a set of prior probabilities for the hypotheses.

Plato used this kind of argument to prove the existence of Atlantis, and philosophers used it to prove the absurdity of Newtonian mechanics.

Clearly the theorem can be extended to any number of hypotheses, the only conditions being that they be mutually exclusive and exhaustive. The technique is useful for guidance only. Except in special cases, it should not be considered as a scientific tool, particularly since the prior probability distribution of the set of hypotheses is required. Typically this is the subject of scientific investigation.

Table of Chi-Square Distribution, 5% Points

Df	Chi-square	Df	Chi-square
1	3.84	19	30.1
2	5.99	20	31.4
3	7.81	21	32.7
4	9.49	22	33.9
5	11.1	23	35.2
6	12.6	24	36.4
7	14.1	25	37.7
8	15.5	26	38.9
9	16.9	27	40.1
10	18.3	28	41.3
11	19.7	29	42.6
12	21.0	30	43.8
13	22.4	40	55.8
14	23.7	50	67.5
15	25.0	60	79.1
16	26.3	70	90.5
17	27.6	80	101.9
18	28.9	90	113.2

The Choice of Computing Software for Log Linear Models

Currently there exists a wide variety of computer programs and packages for analyzing counted data. By far the most commonly used will be the "log linear" capability of the recent issue of the SPSS-X language, and P4F of the BMDP series. Other than availability there is little to choose between these two offerings. Both are very fast and flexible. Since each BMDP program is independent of the others, a smaller computing system is sufficient for this particular series than is true for the larger SPSS-X language. In fact a special microcomputer has recently been advertised specifically for handling the BMDP programs. A version of SPSS, which I have not used, is available for the IBM-PC compatibles. I have also not used SYSTAT, available for the PC, or GLIM each of which will perform all the statistical analyses presented here (Fienberg, 1986).

References

Alexander, R.
   1979   Darwinism and Human Affairs. University of Washington.
Alexander, R.
   1981   "Evolution, culture, and human behavior: some general considerations." In
           R.D. Alexander and D.W. Tinkle (eds.), Natural Selection and Social
           Behavior. Chiron Press.
Alfred, B.
   1970   "Blood pressure changes among male Navaho migrants to an urban en-
           vironment." Canad. Rev. Soc. & Anth. 7:189–200.
Alfred, B.
   1979   "Aspects of the distribution of Hb-S in the United States." Am. J. Phys.
           Anthrop. 52:341–350.
Alfred, B., M. Greig and N. Petrakis
   1979   "Demographic effects on the distribution of some hemoglobin types, G6PD
           (A, B), and Duffy (A, B). Can. Rev. Phys. Anthrop. 1(2):39–45.
Altmann, S. and J. Altmann
   1970   Baboon Ecology. University of Chicago Press.
Armstrong, R.
   1978   "Altruism, group selection, and human evolution." In R.J. Meier, C.M.
           Otten and F. Abdel-Hameed (eds.), Evolutionary Models and Studies in
           Human Diversity. Mouton Publishers.
Atkinson, R., G. Bowers and E. Crothers
   1965   An Introduction to Mathematical Learning Theory. John Wiley & Sons.
Axelrod, R.
   1984   The Evolution of Cooperation. Basic Books.
Axelrod, R. and W. Hamilton
   1981   "The evolution of cooperation." Science 211:1390–1396.
Beall, C. and M. Goldstein
   1981   "Tibetan fraternal polyandry: a test of sociobiological theory." Am. Anthrop.
           83(1):5–13.

Beauchamp, K. and E. Boyse
    1985   "The chemosensory recognition of genetic individuality." Scientific American 253(1): 86–92.
Bishop, Y., S. Fienberg and P. Holland
    1977   Discrete Multivariate Analysis: Theory and Practice. The MIT Press.
Burns, P.
    1979   "Log-linear analysis of dental caries occurrence in four skeletal series." Am. J. Phys. Anthrop. 51: 637–648.
Chagnon, N.
    1979   "Anthropology and the nature of things." In N. Chagnon and W. Irons (ed.), Evolutionary Biology and Human Social Behavior. Duxbury Press.
Clark, C. and M. Mangel
    1984a   "Foraging and flocking strategies: information in an uncertain environment." American Naturalist 123(5): 626–641.
Clark, C. and M. Mangel
    1984b   "The evolutionary advantages of group foraging." Institute of Applied Mathematics, U.B.C. Technical Report No. 84-14.
Cochran, W.G.
    1954   "Some methods for strengthening the common Chi-square tests." Biometrics 10: 417–451.
Daly, M. and M. Wilson
    1983   Sex, Evolution and Behavior, 2nd ed. Willard Grant Press.
Dawkins, R.
    1976   The Selfish Gene. Oxford University Press.
Demetrius, L.
    1974   "Natural selection and age-structured populations." Genetics 79: 535–544.
Demetrius, L.
    1975   "Reproductive strategies and natural selection." American Naturalist 109: 243–249.
Demetrius, L.
    1976   "Measures of variability in age-structured populations." J. Theoretical Biology 63: 397–404.
Dixon, W.L. (ed.)
    1981   BMDP Statistical Software. University of California Press.
Dobzhansky, T. and E. Boesiger
    1983   Human Culture, A Moment in Evolution. Columbia University Press.
Eigen, M. and R. Winkler
    1981   Laws of the Game. Alfred A. Knopf.
Emlen, J.
    1984   Population Biology. Macmillan Publishing Co.
Feller, W.
    1950   An Introduction to Probability Theory and Its Applications, vol. 1. John Wiley & Sons, Inc.
Fienberg, S.
    1977   The Analysis of Cross-Classified Categorical Data. The MIT Press.
Fienberg, S.
    1980   The Analysis of Cross-Classified Categorical Data, 2nd ed. The MIT Press.
Fienberg, S.
    1986   Personal communication.

Flinn, M.
1981   "Uterine vs. agnatic kinship variability and associated cousin marriage preferences: an evolutionary biological analysis." In R.D. Alexander and D.W. Tinkle (eds.), Natural Selection and Social Behavior. Chiron Press.
Fogelin, R.
1982   Understanding Arguments, 2nd ed. Harcourt Brace Jovanovich.
Gatlin, L.L.
1972   Information Theory and the Living System. Columbia University Press.
Gaulin, S. and A. Schlegel
1980   "Paternal confidence and paternal investment: a cross cultural test of a sociobiological hypothesis." Ethology and Sociobiology 1:301–309.
Giere, R.
1984   Understanding Scientific Reasoning, 2nd ed. Holt, Rinehart and Winston.
Goldstein, L. and D. Schneider
1980   Finite Mathematics and Its Applications. Prentice-Hall.
Goody, J. and J. Buckley
1980   "Implications of the sexual divisions of labor in agriculture." In J.C. Mitchell (ed.), Numerical Techniques in Social Anthropology. Institute for the Study of Human Issues.
Haberman, S.
1978   Analysis of Qualitative Data (2 vol.). Academic Press.
Halperin, S.
1979   "Temporary association patterns in free ranging chimpanzees: an assessment of individual grouping preferences." In D.A. Hamburg and E.R. McCown (eds.), The Great Apes. The Benjamin/Cummings Publishing Co.
Hamilton, W.
1964   "The genetical evolution of social behaviour, I & II." J. Theoretical Biology 7:1–16, 17–32.
Hausfater, G.
1975   Dominance and Reproduction in Baboons (Papio cynocephalus). S. Karger.
Hoel, P.
1962   Introduction to Mathematical Statistics. John Wiley & Sons.
Horovitz, A. and T. Ben-Hur
1983   "A model for the kinetics of crime." J. Theor. Biol. 103:609–617.
Hughes, P. and G. Brecht
1975   Vicious Circles and Infinity: An Anthology of Paradoxes. Penguin Books.
Hurwicz, L.
1968   "Game theory and decisions." In D. Messick (ed.), Mathematical Thinking in Behavioral Sciences. W.H. Freeman and Co.
James, W.
1890   The Principles of Psychology. Holt.
Jones, A.
1980   Game Theory: Mathematical Models of Conflict. Halstead Press.
Kac, M.
1968   "Probability." In D. Messick (ed.), Mathematical Thinking in Behavioral Sciences. W.H. Freeman and Co.
Kawanaka, K.
1982   "Further studies on predation by chimpanzees of the Mahale Mountains." Primates 23(3):364–384.

Kemeny, J. and J. Snell
    1960   Finite Markov Chains. D. van Nostrand.
Kemeny, J., J. Snell and G. Thompson
    1966   Introduction to Finite Mathematics. Prentice-Hall, Inc.
Kurland, J.
    1979   "Paternity, mother's brother and human sociality." In N. Chagnon and
            W. Irons (eds.), Evolutionary Biology and Human Social Behavior. Duxbury
            Press.
Landau, B., H. Gleitman and E. Spelke
    1981   "Spatial knowledge and geometric representation in a child blind from
            birth." Science, 213 (11 Sept.): 1275–1277.
Latham, M., R. McGandy, M. McCann and F. Stare
    1970   Scope Manual on Nutrition. Upjohn Co.
Lewin, R.
    1984   "Practice catches theory in kin recognition." Science 223: 1049–1051.
Lewontin, R.
    1974   The Genetic Basis of Evolutionary Change. Columbia University Press.
Lewontin, R. and J. Hubby
    1966   "A molecular approach to the study of genetic heterozygosity in natural
            populations, II. Amount of variation and degree of heterozygosity in natural
            populations of Drosophila pseudoobscura." Genetics 54: 595–609.
Luce, R. and H. Raiffa
    1957   Games and Decisions. John Wiley & Sons.
MacDonald, D.
    1982   "Notes on the size and composition of groups of Proboscis monkey, Na-
            salis larvatus." Folia Primatologica 37: 95–98.
McGraw-Hill Book Co.
    1983   SPSSX User's Guide. McGraw-Hill Book Co.
Maynard-Smith, J.
    1982   Evolution and the Theory of Games. Cambridge University Press.
Maynard-Smith, J. and G. Price
    1973   "The logic of animal conflict." Nature 246: 15–18.
Mednick, S., W.F. Gabrielli, Jr. and B. Hutchings
    1984   "Genetic influences in criminal convictions: evidence from an adoption
            cohort." Science 224: 891–893.
Mill, J.
    1874   A System of Logic, 8th ed. Harper & Row.
Mims, C.
    1970   "Stress in relation to the processes of civilization." In S.V. Boyden (ed.), The
            Impact of Civilization on the Biology of Man. University of Toronto Press.
Muller, H.
    1950   "Our load of mutations." Am. J. Hum. Genet. 2: 111–176.
Murdock, G.
    1967   World Ethnographic Sample. University of Pittsburgh Press.
Nishida, T.
    1979   "The social structure of chimpanzees of the Mahale Mountains." In D.A.
            Hamburg and E.R. McCown, The Great Apes. The Benjamin/Cummings
            Publishing Co.

Olkin, I., L. Gleser, and C. Derman
    1980   Probability Models and Applications. Macmillan Publishing Co.
Packer, C.
    1977   "Reciprocal altruism in *Papio anubis*." Nature 265:441–443.
Parzen, E.
    1960   Modern Probability Theory and Its Applications. John Wiley & Sons, Inc.
Patil, G. and C. Taille
    1982   "Diversity as a concept and its measurement. J. Am. Stat. Assn. 77:548–561.
Peel, R.
    1981   "Natural selection, social evolution and economic strategy." J. Biosocial
            Science 13:377–390.
Polya, G.
    1954   Patterns of Plausible Inference, vol. II. Princeton University Press.
Ransom, T.
    1981   Beach Troop of the Gombe. Bucknell University Press.
Ross, S.
    1972   Introduction to Probability Models. Academic Press.
Rutberg, A.
    1983   "The evolution of monogamy in primates." J. Theor. Biol. 104:93–112.
de Saint-Exupery, A.
    1943   The Little Prince. Harcourt, Brace & World.
Salmon, M.
    1984   Introduction to Logic and Critical Thinking. Harcourt Brace Jovanovich.
Selye, H.
    1950   "Stress." Acta.
Slater, P.J.B.
    1978   "Data collection." in P.W. Colgan (ed.), Quantitative Ethology. John Wiley
            & Sons.
Spencer, P.
    1980   "Polygyny as a measure of social differentiation in Africa." in J.C. Mitchell
            (ed.), Numerical Techniques in Social Anthropology. Institute for the Study
            of Human Issues.
Trivers, R.
    1974   "Parent-offspring conflict." American Zoologist 14:249–264.
US-DHEW
    1980   Selected Genetic Markers of Blood and Secretions. DHEW Publication No.
            (PHS) 80-1664.
de Waal, F.
    1982   Chimpanzee Politics. Harper & Row.
White, H.
    1963   An Anatomy of Kinship. Prentice-Hall, Inc.
Whittaker, J.
    1984   "Stable strategies for evolution." Unpublished.
Wilson, E.
    1975   Sociobiology, the New Synthesis. Belknap.
Woolf, B.
    1955   "On estimating the relation between blood group and disease." Ann. Hum.
            Genet. 19:251–253.

Index